Handbook

# MARINE CORPS INSTITUTE

# IRAQ: AN INTRODUCTION TO THE COUNTRY AND PEOPLE

MARINE BARRACKS
WASHINGTON, DC

# Iraq:

# An Introduction to

# the Country and People

# Introduction

## Overview

**Purpose**

The United States is committed to on-going involvement with Iraq to rebuild the country. Marines, who deploy to Iraq or are involved with Iraqis, need a basic knowledge of the country, its culture, history and present-day state of affairs.

- This handbook explains a number of basic issues that should be in the knowledge "toolbox" of a Marine working with Iraqis or deployed to the region.
- You can use the links that form an integral part of this document to explore further any of the topics discussed in this handbook.

**Contents**

This handbook includes the following topics:



*Continued on next page*

**Contents,**
(continued)



**Three Critical Factors**

There are three important factors that have shaped Iraqi culture and society:

- **Islam** – particularly, the division between Sunnis and Shias (Shi'ites)
- **Oil** – which supplies Iraq with its "life blood"
- **Saddam Hussein** – former leader of the country

You will learn more about each of these as you read this handbook.

*Continued on next page*

# Overview, Continued

**Geography**

***Iraq*** or Irak, officially the Republic of Iraq, a country of 167,924 square miles (434,924 sq km) in southwest Asia is bordered on the

- South by Kuwait, the Persian Gulf, and Saudi Arabia
- West by Jordan and Syria
- North by Turkey
- East by Iran

Baghdad is the capital and largest city. Other large cities are Mosul and Kirkuk and the port of Basra. The country is divided into 18 provinces.

http://www.lib.utexas.edu/maps/iraq.html has a number of good maps showing Iraq.

**Climate and Weather**

Most of Iraq gets little rainfall. The winters are cool; summers are dry and hot, with blowing winds and sandstorms. The exception is the northern mountains along the Iranian and Turkish borders. Here the winters are cold with heavy snows.

Flooding from the rivers because of melting snow is a problem farther south, but is less common today because of flood-control projects. Agriculture depends mainly on water from the rivers.

- Because of the sandy soil and intense heat, Iraqi farmers in the southeast grow "orchards" of dates and cotton.

- Farther up the rivers, as the elevation increases, rainfall allows different kinds of crops, such as grains and vegetables.

- In the mountains of the north, agriculture gives way to oil. The largest oil fields are near Mosul and Kirkuk.

Note: Appendix B provides you with quick facts about Iraq.

(This page intentionally left blank.)

# Iraqi Culture, History, and Religion

## Culture and Peoples

**Cultural and Ethnic Factors**

Iraq with its population of about 22 million people has been shaped by a number of cultural and ethnic factors.

- Most of the people are of Arabic origin. About 80% speak Arabic.
- Ninety-five percent of the people are Muslim. There are about twice as many Shias (Shi'ites) as Sunnis. (Sunnis are the more numerous sect or religious group in most Muslim countries.)
- Kurds are a large minority living in the uplands of northeast Iraq. They are primarily Sunni Muslims.

**Minority Groups**

Other minorities of Iraq include Turks, Armenians, and Assyrians (Nestorian Christians). Most of the country's once large Jewish population resettled in Israel in the early 1950s.

**Urban and Rural**

Iraq's only way of reaching the sea is a short stretch of coast on the northwestern end of the Persian Gulf, including the Shatt al Arab waterway.

For the majority of the population, life centers on the Tigris and the Euphrates rivers, which flow southward and come together in the Shatt al Arab. Urban growth is in this area. There are also numerous marshlands and water basins between these two rivers. The southwest is desert and supports a small population of nomad shepherds. Present day Iraq encompasses the approximate area of the Middle East called Mesopotamia in ancient times.

*Continued on next page*

**Selected Topics** The topics covered in this section on Iraqi history, culture, and peoples are shown in the table below.

# Kurds

**Special Case**

The Kurds are one among a number of minorities living in present-day Iraq. Because of their oppression by the regime of Saddam Hussein, which is covered later in this handbook, you will read about these people separately in this section.

**Ancient History: Key to the Present**

The Kurds are an old, Indo-Iranian tribe who lived in ancient Persian times in areas that the present-day Kurds, descendents of this tribe, still occupy. This area is often called Kurdistan.

**A Distinct Name**

The name *Kurd* was used around 640 B.C. It described a group of nomads that lived on the plateaus and mountains between Armenia and the Zagros Mountains. Later, the Arabs applied the name to the residents of these mountainous areas who had converted to Islam.

**Kurdish Expansion**

- Eventually, the Kurds spread out and settled in a region reaching from central Anatolia (modern-day Turkey) to northeast Iran.
- The Kurds have developed as "isolated" units, physically cut off from one another into what is often called the "Kurdish diaspora."

**Current Population Distribution**

Currently, most of the Kurdish people, numbering about 25 million total, live in Turkey, Iraq, Syria, and Iran, primarily.

- About half of this total population lives in Turkey.
- 6-7 million live in Iran.
- 3.5-4 million, in Iraq.
- 1.5 million live in Syria.

There are groups living outside the Middle East in Armenia, Germany, Sweden, France, and the United States. Kurdish communities also exist in various countries of the former Soviet Union.

**The Kurds in Iraq**

The Iraqi Kurds make up about 23% of Iraq's population. They live near the towns of Mosul, Kirkuk, and Sulaimaniyah. The Iraqi Kurds are sometimes known by the name Failili Kurds.

*Continued on next page*

**People vice Nation**

In spite of the fact that the Kurds as an ethnic group are spread over such a large area, most of them are bound together as a common group by their language, traditions, and their culture. In recent interviews among Kurds living in southeastern Turkey, respondents considered themselves to be Kurds first, Turks second.

The concept of a Kurdish sovereign state is still not a full reality in the minds of all Kurds. It is certainly not an idea that the nation states of the present-day Middle East –Turkey, Iran, and Iraq – want to consider. Their hope is to "integrate" the Kurds into the respective country that they live in.

(See http://ethnicity.your-directory.com/Finding the Kurds a Way Kurdistan and the Discourse of the Nation State'2242643.html for a detailed discussion of the implications of a Kurdish sovereign state.)

**Indo-European, not Arabic**

The Kurds are not an Arabic speaking people. They don't speak a language that is related to either Arabic or Turkish. The Kurdish language can probably be traced back to an ancient language that was more closely related to Farsi (the current language spoken in Iran) and Hindi (the current official language of India). Present-day Kurdish is closely related to the northwestern group of Iranian languages.

**Religion**

The Kurds accepted Islam from the Arabs during the $7^{th}$ century A.D. They are Sunni Muslims, not Shias (Shi'ites).

**Culture**

Kurds have their own music, poetry, and dance. Individual villages have their own dances in which men and women often dance together. Their legends are filled with heroism and romance. Nature, particularly the mountains, has shaped Kurdish life.

- Mountain flowers have contributed to the color and patterning in Kurdish clothing and kilims (woven rugs).
- Men still wear turbans and baggy pants (called shalvari).
- Women wear headscarves, long colorful dresses, and coats decorated with silver and gold thread.

*Continued on next page*

**Economy and Livelihood**

The Kurds' economy has been based on agriculture from ancient times.

- About a third of their land is fit for farming.
- Many Kurds grow wheat and other grains.
- In the higher mountains, the Kurds herd sheep. In the 8th century, Kurdish dairy products were considered so delicious that the Vikings from Scandinavia and medieval Russia traveled south to the Iraqi region to buy butter from Kurdish peasants.

Because of unstable conditions in present-day Iraq agriculture has declined.

**A Distinct Difference**

There is a distinct difference between the Arab-speaking majority of the "desert lowlands" of Iraq and the Kurdish mountain folk.

- Language separates them.
- Culture has developed along separate lines.
- Economically, the Kurds have held to their farming traditions; the Iraqi Arabs have expanded into marketing entrepreneurship.

See http://ethnicity.your-directory.com/Meet_the_Kurds'2242666.html for a good detailed introduction to the culture of the Kurds.

http://www.bartleby.com/65/ku/Kurds.html gives a good overview of Kurds.

http://www.kelim-art.de/exhib/kt/e_kt_index.html is a site that shows the beauty and craftsmanship of Kurdish kilims [rugs].

# Iraqi Economy and Livelihood

**Oil**

The chief factor that dominates Iraq's economy is oil. Traditionally it provided about 95% of the money received from foreign trade. Oil exports have been substantially reduced because of

- The United Nations imposed economic sanctions after Iraq's invasion of Kuwait.
- The destruction of oil fields by Saddam Hussein when the U.S. coalition invaded in 2001.
- Current terrorist sabotage of the oil pipelines.

**Oil Production**

The Iraq Petroleum Company, a national company, used to produce the oil. The oil is pumped to Turkey, Lebanon, Syria, and the Persian Gulf.

- In the early 1980s, oil production averaged about 2.0 million barrels per day (bpd).
- In the late eighties about 1.7 million barrels per day (bpd) were pumped.

**Natural Gas**

Iraq also has reserves of natural gas. In 1987, it produced about 7 million cubic meters. Natural gas reserves are nearly 850 billion cubic meters.

*Continued on next page*

**)ther Goods ınd Services**

Other industries produce chemicals, textiles, cement, food products, construction materials, leather goods, and machinery. Production in electronics, fertilizers, and sugar had recently started. Since the invasion of Iraq the production of these goods and services is at a standstill.

- Services contributed about 40% to the gross domestic product.
- Mining and manufacturing, about 6%.
- Construction employed about 20% of the civilian population and the military forces.

The University of Texas at Austin has good detailed information about Iraq at http://inic.utexas.edu/menic/Countries_and_Regions/Iraq/.

http://www.iraqresearch.com/ is a good Congressional Research Service site that has links about Iraq.

**.abor Force**

Government figures showed the industrial labor force was at 170,000 in 1984, with

- 80 percent of workers in state factories
- 13 percent in the private sector
- 7 percent in the mixed sector

**\griculture**

About 30% of the workforce was usually involved in agriculture.

- The main crops include wheat, barley, rice, vegetables, cotton, and dates (Iraq is one of the world's largest producers).

- Iraqis also raise cattle and sheep.

Agriculture accounts for about 10% of the gross domestic product. Since the downfall of Saddam Hussein agricultural production has fallen off.

**'oreign Dependence**

Iraq depends very heavily on economic aid from both the West and Arab countries. The economic UN embargo just emphasized this. Both exports and imports were significantly reduced under the embargo, causing shortages and a rise in prices in Iraq.

# Religious Division vice Iraqi Nationalism?

**Two Sects**

One of the important factors in understanding Iraq and its people is Islam.

- The main Muslim sects of Islam, Sunnis, and Shias or Shi'ites, dominate the culture of Iraq. The majority in the country are the Shias, but the minority, the Sunnis, ran the country both before and during Saddam Hussein's regime. There was much discrimination against the Shias.
- Most people feel that the Shias would have driven the Sunnis out of power if given the chance. (See appendix A for details on the differences between these two sects of Islam.)

**Current Tension**

One of the key issues today in post-Saddam Hussein Iraq remains the relationship between the majority Shias and the minority Sunnis.

- The Sunnis are now apprehensive that with popular sovereignty, they will lose their political domination of the country.
- How the majority Shias will treat the minority Sunnis is an important issue in post-Saddam Hussein Iraq.

**Historical Division: One Side of the Coin**

There was good historical reason for making the assumption that the Shias would have driven out the Sunnis.

- For many years, the Arabian Sunnis from the areas around Baghdad, Mosul, and Ar Rutbah, the large cities of Iraq, dominated and ruled Iraq.
- Saddam Hussein and his top deputies, all members of the ruling Baath Party, were also Sunnis. Sunnis had also held the top posts in the Iraqi security forces and had been the army corps' commanders.

Most of the Shias live in the south, the most depressed part of Iraq.

*Continued on next page*

**Possible Reverse Side: Nationalism**

However, this historically based idea of "revenge" of the Iraqi Shias may not be as strong as assumed.

- In 1982, in response to aggression by Saddam Hussein, Iran invaded Iraq. Iran, Iraq's neighbor, is primarily a Shia country.

- The expectation of many was that the Shias of Iraq (75% of enlisted Iraqis were Shias) would join with their Shia brothers of Iran and revolt against the Sunnis of Iraq. No such general revolt occurred.

The Shias of Iraq continued to defend their country and the Sunni Baath regime of Saddem Hussein, even after the Iraqi army suffered major setbacks from the Iranian military.

**Exploiting Ancient History**

Perhaps what was at work here was the age-old antagonism between Arabs and Persians.

- The Iraqis, both the Sunnis and the Shias, are an Arabic-speaking people.

- The modern Iranians, while Shias in religion, are descendents of the Persians and speak Farsi, a language totally unrelated to Arabic. Between Arabs and Persians there has always been historically bitter hatred.

Saddam Hussein and his Baath Party capitalized on this ancient enmity. They described the war with Iran in terms of this hatred, characterizing the modern-day conflict as the "Battle of Qadisiyah," a medieval battle in which the Arabs, the true believers bearing Islam, defeated heathen Persia in A.D. 637.

# Another "Division": Secular vice Religious

**Islam vice Baath**

Another cultural division, which has played a role in Iraq in more recent times, is the division between

- the religious believers in the population, whether they are Shias or Sunnis,

and

- the secularists or non-believers, who were primarily of the Baath Party.

**Shia Opposition**

Before the Baath Party, eventually headed by Saddam Hussein, came to power, Shias had made progress in education, business, and law even though they were under-represented in the government.

- Later in the 1980s, Shias gained more representation in the government, proportional to their numbers in the population. For example, of the eight top Iraqi leaders who in early 1988 sat with Saddam Hussein on the Revolutionary Command Council--Iraq's highest governing body-- three were Shias.

- Saddam Hussein further suppressed the Shias so that they had no controlling voice in the Iraqi government.

The question remains if there will be a Shia "revival" in the "new" Iraq.

# Tribes and Villages

**Tribalism**

As with many Middle Eastern countries, primary relationships play a key role in an Iraqi's life. Most family-oriented relations for an Iraqi are rooted in the age-old traditions of tribalism.

This tribalism is based on nomadism, but has been restructured to accommodate the sedentary environment in cities and towns and the limited territory of the modern age, e.g., land ownership.

**Tribal Responsibility**

The primary kinship unit is a tribe or *shaykh*, with the lineage through several generations from the father's side of the family.

- This patrilineal (descent on the father's side of the family) kinship group is responsible for the extended family in feuds, arguments, and "wars."
- It controls marriage and jointly controls parts of the "tribal" land. Mutual assistance is the key to a successful shaykh economy.
- The "elders of the tribe" share authority.
- A shaykh, which has existed through many generations, is usually very centralized and has a developed social hierarchy.

**Clans**

Within the tribe, the primary family unit is the clan, which is made up of two or more lineage groups. Clans are usually part of a particular tribe defined by blood or by adoption.

- The clan supports its members in any dispute with other clans of the same tribe or shaykh.
- Lands belonging to clans of a particular tribe are usually adjacent to one another.

**Villages**

The most widespread social living unit, the village, is usually a mixture of tribal folk, local tradesmen, and government employees. The line between tribal people and village dwellers used to be very distinct. As the government extended social services to the village, the number and influence of non-tribal people grew.

*Continued on next page*

**Role Changing**

Government and service people before and during Saddam Hussein's regime took over the role that tribal leaders and village people used to play. For example, a government school might be set up and competed with a village religious school. Government cooperatives might also work as the "middleman" selling farmer's goods, instead of the farmers doing this themselves.

- In the 1950's most government employees came from Baghdad. With their citified ways, these Sunni bureaucrats often provoked antagonism among the local Shia villagers.

- In the 1980's before the war with Iran, government representatives began to be drawn from the local village population where they were serving.

As you can see, the division between Shia and Sunni also plays a part in the Iraqi bureaucracy.

# Iraqi History: A Rich and Ancient Past

**Introduction**

To get a better picture of the current political situation in Iraq, you will cover a bit about the history of the country. This brief historical survey is outlined as follows:

- "Ancient" Iraq
- Importance of Islam
- British Influence
- Background of the Rise of Saddam Hussein

**Ancient Iraq: First Civilization**

The area that modern Iraq covers is almost the same as that part of the ancient world called Mesopotamia (from the Greek, meaning "between two rivers," in this case the Tigris and the Euphrates). Another name for this region is the Fertile Crescent. Historians often refer to it as the cradle of civilization.

Like Egypt, China, and India, Mesopotamia was one of the ancient centers of civilization. In 4000 B.C., the Sumerians dominated the area.

- These ancient Sumerians built irrigation systems to harness the waters of the Tigris and Euphrates and developed the first cereal agriculture.

- They also created the earliest writing system, called cuneiform, by stamping clay tablets with a reed stylus.

- Banking also originated in ancient Mesopotamia.

**Hammurabi**

- The Sumerians were overrun and conquered by a number of Semitic-speaking tribes. (The Semitic languages include Arabic, Hebrew and a number of ancient, extinct languages.) Thus the Arab-related population became a force in the region early on.

- There was also much bitter infighting between the various tribes indigenous to the area. Finally around 1700 B.C., the region was united under King Hammurabi into one political unit called Babylon or Babylonia.

*Continued on next page*

**Hammurabi's Dynasty: Achievements**

King Hammurabi conquered territory and extended the borders of Babylon. He is remembered for his body of laws, Hammurabi's Code. Its common theme is that the punishment should fit the crime: "An eye for an eye, a tooth for a tooth."

Hammurabi's son, Nebuchadnezzar, designed and had constructed the ancient architectural wonder, the Hanging Gardens of Babylon, about 50 kilometers south of present-day Baghdad.

# Islam: The Crown and Thorn in Iraq's History

---

**Introduction**

While ancient civilization is an important factor in Iraq's past, the conquest and establishment of Islam in the region is **the** historical and religious event that modern-day Iraqis are most proud of.

In A.D. 634, Arab Muslims reached the Euphrates delta. With the fervor of their Islamic jihad, (*Jihad fi Sabeel lillah)* they conquered and converted the Persians there, eventually establishing the Arabic dynasty of the Umayyads in Iraq.

---

**The Caliphate**

To understand the importance of Iraq in the history of Islam and the Middle East, you need to know about the caliphate.

- Early on in the history of Islam after the death of Muhammad, Muslims chose a leader and established a "dynasty" for controlling all Muslims worldwide.
- The caliph was the leader or head of this dynasty, known at the caliphate, which was the administrative, religious, and legal "directorate" of Islam. Later when the Ottoman Turks conquered Iraq and incorporated it into their empire, the caliphate was moved to Turkey.
- The Republic of Turkey abolished the caliphate in the 1920's, but the caliphate as an historical tradition in Islam is very important for modern-day Iraqis.

---

**The Abbasid Caliphate and Baghdad**

Islamic culture and civilization in Iraq and the surrounding Middle East region reached its height under the Arabic dynasty of the Abbacies, who establish the Abbasid caliphate, a successor dynasty to the Umayyads, in Iraq in 750 AD.

The Abbasid caliphate, in what is present-day Iraq, founded the city of Baghdad, which became one of the cultural and economic centers of the Middle Ages. Baghdad was:

- The intellectual center of the world at that time and the cultural capital of the Islamic world.
- A power center where Arabic and Persian culture, intellect, and learning merged to create what all Muslims consider the apogee (high point) of Islam.

---

*Continued on next page*

**Safavids**

Two groups vied for domination in the Iraqi region for the next two centuries. The Safavids were the Persian dynasty that ruled Persia (Iran), Iraq's neighbor. The Safavids:

- Were Shias (Shi'ites) (remember, Shias are one of the two important divisions of Islam).
- Wanted to extend Shia power throughout the Middle East.

The Safavids were very involved in attempting to conquer Iraq in the $16^{th}$ and $17^{th}$ centuries and in establishing Shia as the dominant sect of Islam in the region.

**Ottomans**

The Ottomans, a newly-arrived Turkish tribe, fast gaining ascendancy in Anatolia (modern Turkey) and Asia Minor:

- Were Sunnis (the competing sect of Islam).
- Did not want the Shias, led by the Safavids, to dominate Islam.
- Sought to counteract the Shia aggression of the Safavids.

For the next two hundred year, the Persians (Iranians) and the Ottoman Turks conquered and re-conquered various parts of the Iraqi region. The Ottoman Turks eventually dominated Iraq and incorporated it into their empire.

(Good information on the early history of Iraq can be found at the web site http://home.achilles.net/~sal/iraq_history.html .)

*Continued on next page*

**Iraqi Nationalism**

One of the important legacies of the Ottoman domination of the Iraqis was the beginning of Iraqi nationalism.

- Copying a group of Ottoman Turkish nationalists called the Young Turks, Iraqi intellectuals founded an Iraqi nationalist society called *Al Ahd* (the Covenant) in the early 1900s.
- The aim of this group was to gain more autonomy for Iraqis from the Ottomans.

Offshoots of *Al Ahd* formed "cells" among Iraqi officers serving in the Ottoman army and in larger towns such as Mosul and Baghdad.

(For more information on this period see the following web site: http://memory.loc.gov/cgi-bin/query/r?frd/cstdy:@field(DOCID+iq0018 )

## British Influence

**Post World War I**

The British invaded Iraq as part of their campaign against the Ottoman Empire during World War I. After the war, the Treaty of Sevres gave Great Britain control of Iraq as a mandate of the League of Nations.

- In 1921 the country was given a monarchy headed by King Faisal I.
- By 1926, an Iraqi parliament was governing the country although the British, under a 1924 treaty, had a veto over legislation.
- The British mandate ended in 1932 when Iraq was admitted to the League of Nations.
- Iraqi oil exports began in 1934.
- During the period 1936-41, a time of instability in the country, Iraq experienced seven military coups.

(For a fascinating narrative on how the British set up the borders of Saudi Arabia, Kuwait, and Iraq in 1922, go to the web site http://www.nybooks.com/articles/3032.)

**World War II**

Initially, Iraq sided with Germany and Italy in World War II, but after the pro-Axis Iraqi leader, al-Gaylani, was defeated by the British in early 1943, the country joined the Allies and declared war on Germany.

# Rise of Saddam Hussein: Background

**Key Factors**

To understand the rise of Saddam Hussein to power in Iraq, you need to know about certain key facts in Iraqi political history:

- Constant conflict including coup and counter-coup has been a common feature of Iraqi political life for many years.
- The relationship of Iraq with its neighbors, particularly Iran, played a role in influencing Saddam Hussein.
- The growth of the Baath Socialist Party was important too.

We will consider this context in discussing Saddam Hussein's emergence as Iraq's leader.

(This is an excellent article on the family "clan" of Saddam Hussein: http://www.meforum.org/article/273 .)

**Continual Conflict**

The period of the fifties, sixties, and seventies was one of continual conflict in Iraq:

- In the late 1950s the monarchy was overthrown.
- During the 1960s the Baath Socialist Party under Hasan al-Bakr came to power in the country. Coup and counter coup with espionage trials were common during this time.
- The Baath-led government was overthrown in a coup led by President Abdul Salam Mohammad Arif, who was succeeded by his brother, General Abdul Rahman Arif.
- In July 1968, the Baath Party regained power in a coup under Ahmed Hasan al-Bakr, who then became president of the country.

*Continued on next page*

**Iraqi Foreign Relations**

To understand some of the reasons for Saddam Hussein's aggression toward his neighbors, you need to know a bit about Iraq's relations with neighboring countries before Saddam's rise to power in the country.

- Iraq participated in the Arab-Israel War of 1973 and in the oil boycott against nations supporting Israel.
- In early 1974, continual border clashes with Iran erupted in an armed conflict along the entire Iraq-Iran border.
- In 1975 the Kurds, supported by the Iranians, again revolted in northern Iraq, fighting for their independence. Iraq bombed parts of Iran, which created more tension between the two countries.
- In March 1975, Iraq and Iran signed the Algiers Agreement, which defined the boundary of the Shatt el-Arab waterway. Iran then withdrew support from the Kurds, who suffered enormously with intense Iraqi bombings.
- The Islamic revolution in neighboring Iran was recognized as a threat to secular Baath party control of Iraq and caused great anxiety in Iraq.

As you can see, military aggression has played a role in Iraq's relations with its neighbors!

Detailed information on this period and on the rise of Saddam Hussein is at

http://memory.loc.gov/cgi-bin/query/r?frd/cstdy:@field(DOCID+iq0023)

http://memory.loc.gov/cgi-bin/query/r?frd/cstdy:@field(DOCID+iq0022).

# Rise of Saddam Hussein: Baath Party

**Early History**

The Iraqi Baath Party was founded in the 1940s by two Syrian students.

- The word *Ba'ath* or *Baath* means resurrection in Arabic; hence, the party's goals were to "resurrect" the Arabs of Iraq as an ethnic group through socialism, freedom, and unity. This was supposed to occur in Iraq the way it was happening in neighboring Syria.

- The party grew quickly in the early 1950s. The party had difficulty after an assassination attempt in the late fifties.

**One Party State: Saadi**

Saadi, one of the chief leaders of the party, was the Baath political leader who consolidated the party's power. Although the Baathists established the National Council of Revolutionary Command, which appointed the prime minister and the president, Saadi was the real power behind the government. In other words, the party controlled the state.

What the party needed in its role as the nation's leader was cohesion. Saadi gave it this cohesion. He established a one-party state that tolerated little opposition.

**Coup and Counter Coup**

The Baath-led government was overthrown in late 1963, only to regain power in 1968.

The Baath Party of 1968 was much better organized and was a more cohesive unit than in 1963. The party also was aided by its reorganization.

- The party gained broader support based on its newly established militia and intelligence system.

- It also had local party branches that supported the central Baath Party.

*Continued on next page*

---

**The Tikrit Connection**

Tribal ties bound ruling members of the Baath Party together.

- Most of them were Tikritis, Sunni Iraqis from Tikrit, a town in northwestern Iraq.

- Some of the party members were also related to the new president, al Bakr. Besides the prime minister, who was Tikriti, Saddam Hussein himself was also related to al Bakr and from a town near Tikrit.

The one-party system was thus reinforced by this coterie of Tikritis, who wielded the political power in the country.

More information on Saddam Hussein can be found at:

http://www.bartleby.com/65/hu/HusseinS.html

http://home.achilles.net/~sal/iraq_history.html

---

# Saddam Hussein, Former Leader of Iraq

**Early Years**

Saddam Hussein had been a long-standing member of the Baath Party of Iraq. He fled Iraq in 1959 after taking part in an assassination attempt against then Prime Minister Kaseem. He then lived in Egypt and attended law school there.

- Saddam Hussein returned to Iraq in 1963 when the Baathists came back to power for a short time.

- Later he played a leading role in the 1968 revolution that gave the Baath Party control of the country, supporting Ahmed Hasan al-Bakr, who became the Baath chairman of the Revolutionary Command Council and the President of Iraq.

**The Baathist Tikrit Team**

President al Bakr and Saddam Hussein, both Tikritis, dominated the Baath Party, which returned to power in 1968.

- Al Bakr had espoused pan-Arab causes and gave the party its legitimacy. He also ensured support from the army among Baathist and non-Baathist officers.

- Saddam Hussein was **the** politician. His expertise was in organizing secret opposition operations. He knew how to outmaneuver and eliminate political opponents when necessary.

As time went on, it became clear that Saddam Hussein was the moving force behind the party, the political aficionado who worked behind the scene. He personally

- Directed Iraqi efforts to deal with the Kurdish problem
- Organized the party's infrastructure

*Continued on next page*

**Saddam Hussein -- alone**

In the mid-seventies, Saddam Hussein began to consolidate his political power.

- With President al Bakr in failing health, most of the ministers, who should have been reporting to the president, reported directly to Saddam Hussein.
- Saddam Hussein showed his modus operandi: He did not share power and considered the cabinet and the council rubber stamps for his will.

In July 1979, President al Bakr resigned, and Saddam Hussein officially became President of the Republic, Secretary General of the Iraqi Baath Party Regional Command, Chairman of the Regional Command Council, and Head of the Iraqi Armed Forces.

# Saddam Hussein's Policies

## ntroduction

**'urpose**

Rather than "narrate" the history of Iraq under Saddam Hussein's dictatorship, we will choose key topics that show his policies. The topics discussed show his domination and oppression.

**;elected Topics**

The topics covered in this section are shown in the table below.

# Kurdish Oppression

**Early Oppression and Resistance**

Kurdish history, whether in Turkey, Iran or Iraq, has been a tradition of oppression with the aim of controlling the Kurdish population. This has resulted in strong military resistance from the Kurdish side. The Iraqi government has characterized the Kurdish response as rebellion and revolt. In Iraq the oppression and its response is outlined thus:

- In 1958 after the overthrow of the Iraqi monarchy and the establishment of the Baathist-led government, the Kurds hoped for more autonomy in administration. The new Iraqi government failed to follow through.

- During the 1960s, Iraqi Kurds under the leadership of Mustafa al-Barzani waged a lengthy military campaign with the aim of establishing a unified and autonomous Kurdistan in Iraq.

**1970's**

In 1970 the central government promised local self-rule to the Kurds. Their capital was to be Erbil. The Kurds rejected the proposal for two reasons:

- The proposed autonomy did not really give Kurds authority. The central government still retained control.

- As part of the newly proposed Kurdish autonomous region, the Kurds were not granted control of Kirkuk, an important city in the Kurdish area and a key oil-producing center.

Heavy fighting between the central government and Kurdish forces erupted when the central government tried to force limited autonomy on the Kurds in 1974.

*Continued on next page*

**Saddam Hussein's "Campaign"**

In 1979, the Baathist-led government of the new Republic of Iraq launched a campaign of terror and murder against the Kurds. It also carried out a program of systematic assassination of Kurdish leaders.

Saddam Hussein's reign of terror continued for eight years, culminating in 1987-1988 with the Anfal Campaign.

- His forces tortured and killed Kurds.
- He destroyed many Kurdish villages, forcing the surviving villagers to live in special zones and camps where he could control them.

Finally, Saddam Hussein ordered the use of poison gas attacks on Kurdish villages and the arrest and execution of Kurdish males in a final effort to root out all resistance. The result that year was the murder of 200,000 people.

**Post-1991 Persian Gulf War**

In 1991 after the end of the Persian Gulf War, Iraqi forces again crushed another Kurdish uprising. Fearing gas attacks again, nearly 500,000 refugee Kurds attempted to flee to Turkey, and one million fled over the border into Iran.

Later when an autonomous U.N.-protected region for Iraqi Kurds was established in northern Iraq, thousand of Kurds returned to and resettled in Iraq. A general election was held in the autonomous region.

**Kurdish "Politics"**

The Kurds themselves have internal problems. They are split into two opposing factions, the Kurdistan Democratic Party and the Patriotic Union of Kurdistan. In 1999 these two groups ended hostilities.

# Internal Control

**Cult Atmosphere**

Saddam Hussein created a personality cult that fed on his will, totally. He effectively silenced all opposition to his regime in Iraq. For over 20 years, his style of oppression ruled Iraq, silencing a people that had once been a rich source of thought and culture in the Middle East.

**Suppression of Human Rights**

- Saddam Hussein did not allow his people to vote.
- Iraqis had no freedom of expression. The daily Iraqi newspaper was owned by his son.
- There was no freedom of association and movement in the country. Iraqis, allegedly, could only assemble to show their support for Saddam Hussein and could only vote for him in elections. Iraqis were not free to leave the country.
- Saddam Hussein established a large network of spies, through blackmail and fear. In some instances, members of a family were so terrorized that they reported on their relatives.

*Continued on next page*

---

**Regime of Systematic Torture and Terror**

The following allegations have been made against the regime of Saddam Hussein:

- Collective torture -- Saddam Hussein used collective torture to suppress dissent and to instill fear in anyone considering opposing his will. Whole families or ethnic groups (the Kurds were a good example) were tortured for the actions of one dissident.

- Public torture and humiliation -- Saddam Hussein's forces respected no person's dignity. Women were allegedly raped and tortured in front of their families, many were videotaped in order to blackmail their families. Suspect citizens were cruelly dismembered; their families had to pay for the instruments of death and then display the body as a warning to others who might be considering expressing an opinion.

- Branding and Amputation -- Hospitals were known to brand dissidents or to amputate their limbs. In 1994, Saddam Hussein issued a number of decrees establishing cruel and inhuman penalties such as amputations. Military deserters were punished by amputation.

- Chemical weapons -- Saddam Hussein also used chemical weapons against his own people. He decimated more than 60 villages and murdered 30,000+ Iraqis with poison gas.

- Summary executions -- There were numerous ways that Saddam Hussein's forces did this. One "efficient" method was to line up all the males of a village and shoot them one at a time. But much of the time, a slow, excruciating death was preferred. Political prisoners were given slow-acting thallium, a poison that "permeates" the human body and takes several days to bring death. In many instances, the bodies were buried in unmarked sites.

---

*Continued on next page*

**Religious and Social Atrocities**

Besides oppressing the Kurds, the largest ethnic minority in the country, Saddam Hussein tried to silence other ethnic and religious groups.

- He waged terror and oppression against religious believers of Iraq.
- Through murder, torture, and limitations on their religious activities, he suppressed many clerics and scholars in the country, particularly the Shia Muslims.

Anyone whom Saddam Hussein feared would oppose his will was subjected to torture and arbitrary death. This began almost immediately after he became president in July 1979. Claiming "traitors" in the Iraqi National Assembly, he ordered his security forces to publicly and forcibly imprison and eventually kill several leading members of the Iraqi National Assembly.

**Children: Pawns of Saddam Hussein**

Children, allegedly, were not just careless "by-stander" victims in Saddam Hussein's reign of terror and suppression.

- Pharmaceutical supplies for sick children were supposedly exported for resale overseas. Medical supplies were often delayed because the regime demanded bribes from the suppliers. Poor health care for children led to the reemergence of diseases such as polio and cholera in Iraq.
- Saddam Hussein's forces supposedly abducted and held minority children hostage to force their families to relocate in particular regions, especially those families near oil-producing areas.
- There were also weapons training courses for teenagers, subjecting them to 14 hours of physical training and psychological stress each day.
- In many cases, children were allegedly forced to become spies, reporting on their parents' and relatives' conversations and activities to Saddam Hussein's internal security organization.

(See the site http://www.state.gov/g/drl/rls/15996.htm for details on how Saddam Hussein is silencing and controls his own people.)

# nternational Response to Saddam Hussein's Oppression

**Jnited Nations Resolutions**

In Security Council Resolution 677, November 28, 1990, the United Nations condemned Saddam Hussein for forcibly resettling various groups in order to alter the population in different areas of Iraq.

In Security Council Resolution 688, 5 April 1991, the United Nations condemned the repression of Iraqi civilians.

**Jnited Nations Reports**

After a 1999 trip to Iraq, the United Nations special representative on human rights reported he had personally received reports and testimonies of

- People who showed him their scars from government torture
- Women whose family members had been abducted by Iraqi authorities and then had disappeared.

He felt there was a prevailing regime of systematic human rights violations in the country.

In the United Nations Secretary General's Report of 2001, the U.N. special representative stated,

> *"The mere suggestion that someone is not a supporter of the President* [Saddam Hussein] *carries the prospect of the death penalty."*

# Iran-Iraq War, External Domination

**Background**

- As discussed earlier, the antagonism between Iran and Iraq is age-old. It is based partly on the clash of two different cultures: one, Arab-based; the other, Persian.

- Add to this, the religious competition between the two prominent Islamic sects, the Sunnis, who control Iraq, and the Shias, who control Iran, and there is bound to be constant tension between these two giants of Southwest Asia.

**Political Aims**

Saddam Hussein's aim at domination focused on this constant friction point between Iraq and Iran. His foreign policy emphasized

- Strengthening Iraq's role in world oil production
- Gaining the leading role in the Arab world and the Middle East

Accomplishing these two aims had to come at the expense of Iran, which under both the former Shah and the later Islamic religious regime of Ayatollah Khomeini had also sought to expand its influence in these two areas.

**Causes of the War**

- In 1980, Saddam Hussein escalated a dispute with Iran over control of the Shatt al Arab waterway by attacking Iran.

- Besides the territorial claim over the waterway, Saddam Hussein also maintained that Iran had been carrying out artillery attacks on Iraqi territory.

*Continued on next page*

# Iran-Iraq War, External Domination, Continued

**Territorial Aims**

Saddam Hussein wanted to

- Reclaim the Shatt al Arab waterway for Iraq
- Seize the western region of Khuzestan in Iran, an area of extensive oil fields

**Western Involvement**

- France, Germany, and the Soviet Union were the main arms suppliers to Iraq. The United States and Great Britain also quietly provided arms and related military assistance.

- Iran received funding from some Arab countries.

**Initial Success; Later Defeat**

The Iraqis were successful at first in their aggression. They captured the Iranian port of Khorramshahr by late 1980; however, Iranian resistance strengthened so that Iraq had to withdraw in 1982.

Ayatollah Khomeini, the leader of Iran, declared that his country would only stop fighting when Saddam Hussein's government had been destroyed. Iran then began a series of successful assaults that caused Saddam Hussein to use poison gas against Iranian troops.

- Iran captured and occupied some oil-rich islands from Iraq in 1984.
- Both countries began to attack each other's capital cities in 1985.
- Iran captured part of Iraq's Fao peninsula in 1986.

**Ending the War**

- After Iran began to attack Kuwaiti oil tankers in the Persian Gulf, the United States and some European nations got involved in the struggle.

- Attacking a neutral nation like Kuwait caused problems for Iran. The country was not able to buy enough weapons to continue the war.

Finally, in July 1988, Iran accepted a United Nations cease-fire.

**Summary**

Saddam Hussein's principal aim in the Iran-Iraq War was to extend his power in the Middle East. He wanted to dominate the area economically in terms of oil production and militarily by acquiring Iranian territory. He didn't succeed.

# Persian Gulf War

**Causes of the Kuwaiti Invasion**

Saddam Hussein's invasion of Kuwait, which was the ultimate reason for the Persian Gulf War against Iraq, was the result of a territorial dispute. Saddam Hussein accused Kuwait of the following:

- Violating the Iraqi border to get oil resources from the disputed Rumaila oil field
- Flooding the world market with Kuwaiti oil, causing a decrease in oil prices

**Failed Negotiations**

- After a number of failed meetings between Kuwait and Iraq, sponsored by Arab mediators
- After assurances of the United States' noninvolvement in the dispute

Saddam Hussein's forces invaded and occupied Kuwait on August 2, 1990, later annexing the country as the nineteenth province of Iraq.

**United Nations Involvement**

The United Nations Security Council and the Arab League condemned the invasion. Acting under Articles 39 and 40 of the United Nations Charter, the United Nations passed Resolution 660, which

- Condemned the Iraqi invasion of Kuwait
- Demanded immediate withdrawal of Iraqi forces from Kuwait.

**The Embargo and Further UN Resolutions**

Four days later, the United Nations passed an economic embargo that prohibited nearly all trade with Iraq. There followed a series of United Nations resolutions including:

- Further condemning the Iraqi invasion
- Demanding again the withdrawal of Iraqi forces
- Expressing sympathy for and grave concern over the treatment of Kuwaiti nationals in occupied Kuwait
- Stressing the importance of the humane treatment of Kuwaitis under terms of the Geneva Convention
- Laying out specifics of the international embargo against Iraq

*Continued on next page*

---

**Resolution 678**

This series of U.N. resolutions culminated in United Nations Security Council Resolution 678: It allowed U.N. member countries to use all necessary means, including military action, against Iraqi forces occupying Kuwait and demanded total withdrawal of Iraqi forces by January 15, 1991.

---

**United Nations Coalition War**

Operation *Desert Shield*, preparations and the build-up for the eventual invasion of Iraq, began in late 1990.

On January 17, 1991, the U.N. allied coalition of 28 countries, led by the United States, launched aerial attacks on Baghdad in Operation *Desert Storm*, popularly known as the Gulf War.

- The war lasted six weeks. Over 140,000 tons of munitions were unleashed over Iraq. About 100,000 Iraqi soldiers were killed. Bombing destroyed the Iraqi infrastructure including the oil refinery system. All services were totally disrupted.
- On February 24, 1991, the U.S.-led coalition forces entered Iraq with ground forces as part of the plan for liberating Kuwait, but did not penetrate as far as the Iraqi capital Baghdad.

---

**Conclusion to Hostilities**

- The United States announced a cease-fire on February 28, 1991.
- In April 1991, Iraq agreed to a permanent cease-fire with strict conditions.
- Saddam Hussein still remained in power as the head of the Iraqi government.

This site discusses the reconstruction of Kuwait after Saddam Hussein's defeat and the withdrawal of Iraqi forces: http://www.mepc.org/public_asp/journal_vol7/0006_tetreault.asp .

---

# Defiance of United Nations: Restrictions on Iraq

**Continued Embargo**

United Nations Resolution 687 began the cease-fire, created the United Nations Special Commission on Weapons and continued the embargo tying it to Iraqi weapons. The sanctions sought to

- Eliminate Iraq's weapons of mass destruction and delivery systems
- Return to Kuwait all prisoners of war and property taken during the Gulf War
- Establish the principle of Iraqi compensation for war damages

Further the sanctions

- Insisted that Iraq honor its international debts
- Demanded that Iraq stop all terrorism

A Goods Review List (GRL) listed all those items under embargo.

**More Minority Persecution**

One of the key problems in Iraq after the United Nations cease-fire was Saddam Hussein's continued oppression of the Kurds of northern Iraq. Iraqi aircraft and troops were sent there to regain control of the cities that the Kurdish resistance groups had taken during the Gulf War.

Fearing a repeat of the gas attacks of 1988, an exodus of Iraqi Kurdish refugees (about 2.5 million in total) massed on the Turkish and Iranian borders. In the south, Saddam Hussein's forces attempted to control the marsh Arabs, many of whom are Shias.

- United Nations Resolution 688 condemned this continued repression of the Iraqi civilian population.
- Operation *Provide Comfort* established a United Nations protected autonomous region in most of the Kurdish areas of Iraq.

*Continued on next page*

**Exclusion Zones**

To prevent the genocide of the Kurds in the north and the marsh Arabs in the south, the United States, together with France and Great Britain, established air exclusion zones or no fly zones north of the 36th parallel and south of the 32nd parallel, areas prohibited to Iraqi forces.

- The Iraqi air force was not allowed into these exclusion zones. These no-fly zones allowed Kurdish refugees to return home and live without fear of Iraqi reprisal in their autonomous region as allied planes continually surveyed and guarded these exclusion regions.

- As of 1 October 2000, it was reported that in 16,000 sorties since the beginning of 1997, air force pilots had launched more than 1,000 bombs and missiles against 250 targets in northern Iraq in protecting that no fly zone. (See Thomas E. Ricks, "Containing Iraq: A Forgotten War," *Washington Post*, October 25, 2000.)

There was also a no drive zone in southern Iraq to prevent Saddam Hussein from again massing forces on the Kuwait-Iraq border.

---

**Continued Incidents**

During the Clinton administration, Iraq attempted to assassinate former President George Bush while he was visiting Kuwait. The United States response was to fire cruise missiles at Iraqi intelligence headquarters in Baghdad.

In October 1994, Saddam Hussein's republican guard moved toward Kuwait. The United States responded with troop deployments to the Persian Gulf. Saddam Hussein backed down and agreed to recognize the existence and borders of Kuwait.

---

**Sanctions Modifications**

The original economic sanctions imposed on Iraq had been total. Because of humanitarian needs of the Iraqi civilian population, the United Nations modified the sanctions by establishing the Oil for Food Program in 1995 in United Nations Security Council Resolution 986.

- This resolution allowed Iraq to sell a billion dollars of oil to purchase humanitarian supplies, primarily food and medicine, and supplies necessary for the civilian infrastructure of the country.

- The first shipments under this program began arriving in Iraq in 1997.

---

*Continued on next page*

# Defiance of United Nations: Restrictions on Iraq, Continued

**Sanctions Modifications,** (continued)

- Under this program, Iraq could sell oil for humanitarian needs.
- The United Nations controlled all revenues from such oil sales, and the contracts to implement this program were subject to United Nations oversight.
- The program was extended and modified through the years.

**A Slow Process**

The process of "food" for oil was slow.

- Some funds from oil sales might have been diverted to purchase non-humanitarian supplies.
- Contract reviews were tedious.

Critics of the program called for revised sanctions that were more targeted and credible.

The U.S. Department of State web site has a myth/facts sheet about the effect of sanctions on Iraq at http://www.state.gov/p/nea/rls/01fs/3935.htm .

http://home.achilles.net/~sal/un-ros.html has a list of the United Nations resolutions on sanctions.

# Arms Inspections and Disarmament

**Purpose**

- The principal purpose of United Nations Resolution 687 was to eliminate Iraq's capability to inflict damage and make war on its neighbors. Essentially the resolution required the elimination of Iraq's weapons of mass destruction and delivery systems.

- The resolution also charged Iraq to "forego the future development or acquisition of weapons of mass destruction (WMD)."

**Definition**

Four categories of WMD, for declaring, destroying, or rendering harmless, were defined in the resolution:

- Nuclear weapons
- Chemical weapons
- Biological weapons
- Missile delivery systems

**Implementation**

To ensure that Iraq followed the resolution by declaring, destroying or rendering harmless its WMD, the United Nations:

- Established The United Nations Special Committee (UNSCOM).

- Charged UNSCOM and the International Atomic Energy Agency (IAEA) to inspect Iraq to ensure its compliance with the United Nations resolution on declaring, destroying, and rendering harmless Iraq's WMD.

## Inspections, 1991-1998

**Iraqi Obstructionism**

From the start of the UNSCOM inspections, as they were called, Saddam Hussein had been determined to keep a sizeable part of his WMD arsenal.

- This led to constant dissembling and obstruction on the part of the Iraqi government during UNSCOM inspections.
- Saddam Hussein's security force coordinated a campaign of concealment and deception, including hiding documents related to WMD and materiel. This made it extremely difficult to resolve, let alone inspect, issues and materiel related to Iraq's WMD program.
- Iraqi officials outwardly maintained a façade of cooperation, while at the same time delaying or denying UNSCOM inspectors access to facilities, documents, and personnel in a purposed effort to deny information to UNSCOM on Iraq's WMD program.

**United Nations Response**

Prompted by this Iraqi obstructionism, the United Nations Security Council passed several resolutions demanding that the Iraqi government

- Cooperate with UNSCOM and IAEA in the inspection process
- Provide UNSCOM and IAEA with " immediate and unrestricted access to any site they wished to inspect"

**The Iraqi Arsenal**

The coalition military strikes during the 1991 Persian Gulf War and UNSCOM inspections through 1998 destroyed most of Iraq's prohibited ballistic missiles and some Gulf war-era chemical and biological munitions. Iraq still had

- A small force of extended-range Scud missiles
- Some pre-chemical weapons
- Stocks from which biological weapons could be produced
- Munitions suitable for chemical and biological agents

*Continued on next page*

**The Iraqi Arsenal,** (continued)

Following the Gulf War, Iraq may have also

- Preserved and enhanced the infrastructure and expertise necessary for WMD production
- Maintained a stockpile of WMD and increased its size and sophistication in some areas

**Hiatus in United Nations Weapons Inspections**

In December 1998, the Iraqi government refused to continue working with UNSCOM.

- Onsite inspections by UNSCOM were prohibited.
- Technical monitoring systems installed by the United Nations at known and suspected WMD and missile facilities did not operate.
- Iraq prohibited Security Council-mandated monitoring over-flights of Iraqi facilities by United Nations aircraft and helicopters.
- Iraq had also stopped most IAEA inspections since 1998.

After the United Nations staff had been evacuated from Baghdad, the United States and Great Britain launched a bombing campaign, Operation *Desert Fox*, to destroy Iraq's nuclear, chemical, and biological weapons programs.

http://www.cia.gov/cia/publications/iraq_wmd/Iraq_Oct_2002.htm is a detailed discussion of Iraq's WMD and the UNSCOM effort, particularly through 1998.

# Important Post-1998 Events

**Terrorism: The Focus**

The terrorist attacks on the World Trade Center in Manhattan and the Pentagon, with its connection to Al Qaida-sponsored terrorism, brought Saddam Hussein and Iraqi armaments to the forefront of the Bush administration's policy on terrorism.

Assertions by President Bush and his administration now focused on Saddam Hussein as the primary terrorist security risk to the United States and the world, emphasizing the following:

- A connection existed between Saddam Hussein and terrorism by his giving "aid and comfort" and training to such terrorists.
- An axis of evil, including Saddam Hussein, is a concerted effort at state-sponsored terrorism.
- Saddam Hussein had never met his disarmament obligations under numerous United Nations resolutions and implementation inspections.
- There was a dangerous possibility that Saddam Hussein might supply terrorists with WMD, particularly biological weapons, such as smallpox.

**Heightened U.S./U.N. Actions**

In September 2002 President Bush urged the United Nations to encourage Saddam Hussein to comply with United Nations resolutions or "actions will be unavoidable."

- Bush said that Saddam Hussein had repeatedly violated 16 United Nations Security Council resolutions, which included a call for Iraq to "*disarm its chemical, biological, and nuclear weapons programs.*"
- Iraqi officials rejected Bush's assertions.

In response to the United States' heightened effort and possible direct action against Iraq, the United Nations passed resolution 1441. (See the next section for provisions of this resolution.) In mid-November, Iraq accepted the return of United Nations weapons inspectors under resolution 1441.

(See the UN site, http://www.un.org/apps/news/storyAr.asp?NewsID=5344&Cr=iraq&Cr1= for information on this acceptance.)

*Continued on next page*

**Jnited Nations ;ecurity Council Resolution 1441: mportant 'eatures**

The key provisions of Resolution 1441 were that Iraq must:

- Submit to the United Nations Monitoring, Verification and Inspection Commission (UNMOVIC), the IAEA, and the Security Council a complete accounting of all its nuclear, biological, and chemical programs, all ballistics missiles and other delivery systems, with a precise listing of all weapons locations, stocks of agents, and all production facilities.

- Provide UNMOVIC and IAEA with "unimpeded, unconditional, and unrestricted access" to all types of facilities and transport which the two agencies may want to inspect.

- Allow unrestricted interviews to all personnel that UNMOVIC and IAEA may want to talk with, either inside or outside Iraq. The families of interviewees may also travel with the interviewees. Interviews may occur without the presence of Iraqi government observers.

If Iraq failed to comply with this resolution, it would be in violation of the resolution, which would be reported to the Security Council. The resolution said nothing about the "punishment" that Iraq would incur if it didn't fulfill this resolution.

A complete text of the resolution is at: http://www.un.org/Docs/journal/asp/ws.asp?m=S/RES/1441(2002) or at http://ods-dds-ny.un.org/doc/UNDOC/GEN/N02/682/26/PDF/N0268226.pdf?OpenElement

**The Mandate**

United Nations Security Council Resolution 1284 created the United Nations Monitoring, Verification and Inspection Commission (UNMOVIC), which replaced UNSCOM. The mandate of UNMOVIC was to

- Continue with UNSCOM's efforts to disarm Iraq of its weapons of mass destruction

- Check and monitor Iraq's compliance with its obligations not to reacquire the weapons prohibited by the United Nations Security Council

(The United Nations site http://www.un.org/apps/news/infocusRel.asp?infocusID=50&Body=Iraq&Body1=inspect has information on the resolutions and the inspections.)

# Iraq's Weapons Declaration

**Compliance**

Iraq allowed United Nations inspectors back into the country. Cooperation with U.N. personnel, however, was not full or complete.

- The Iraqi government did not allow personnel associated with its various WMD programs or facilities, such as scientists and technicians, to talk freely with UNMOVIC inspectors. An Iraqi security official was invariably present.

- Iraq submitted its required weapons declaration to the United Nations, but the report had irregularities and gaps, some of which are discussed below.

**Biological Agents**

In the declaration on its weapons of mass destruction to the United Nations, Iraq omitted a number of key biological agent items, including:

- 2,160 kilograms of growth media for anthrax
- 1,200 liters of botulism toxin
- 5,500 liters of clostridium

The United States asked why the Iraqi government ignored these agents in its declaration.

**Ballistic Missiles**

- Iraq said it was manufacturing new energy fuels suited for a class of ballistic missile which it had not admitted it had previously possessed.

- The United States government asked why Iraq was manufacturing fuel cells for missiles it said it didn't have.

*Continued on next page*

**Chemical & Biological Weapons Munitions**

In January 1999, the United Nations Special Commission reported that Iraq had **not** provided credible evidence that:

- 550 mustard gas-filled artillery shells and 400 biological weapon-capable aerial bombs had been lost or destroyed.
- The Iraqi government had not adequately accounted for these chemical weapons in the declaration.

The U.S. government asked what the Iraqi government was concealing by not including these munitions in its declaration.

**Mobile Biological Weapon Facilities**

The Iraqi declaration provided no information about its mobile biological weapon agent facilities, i.e., biological labs in trucks. Instead it insisted that these were "refrigeration vehicles and food testing laboratories."

Again, the United States asked what was the Iraqi government concealing about these mobile facilities.

More information about the declaration is at
http://www.state.gov/r/pa/prs/ps/2002/16118.htm

## A New Coalition

**The Smoking Gun**

Many members of the United Nations were looking for direct evidence of Iraq's WMD before considering the possibility of any intervention in the country.

- Russia, China, France, and Germany were against immediate military intervention in Iraq.
- The reluctance of other United Nations members made it difficult for the United States to build an allied coalition for invading Iraq.

**Beginning a Coalition**

The United States was working on building a new coalition for the possible attack on Iraq. Many nations that participated in the 1991 Persian Gulf War were reluctant to join the coalition.

- France, a permanent member of the Security Council, had not ruled out military force against Iraq, but said that United Nations inspectors must be given the broadest opportunity to fully do their job; otherwise, France would veto any United Nations resolution authorizing force against Iraq.
- Germany said it would not support a military campaign.
- Popular opinion in much of the world was against military intervention without United Nations approval.

**Turkey**

The Republic of Turkey is an example of a reluctant coalition partner. Turkey is a non-Arabic secular state.

- It had to be one of the key countries in any coalition strategy against Iraq because part of the Bush administration's military strategy was to attack Iraq from the north.
- The northern prong of this possible invasion would come from Turkey.

*Continued on next page*

**Turkish Apprehension**

- Most Turks are Muslims like the Iraqis.
- Additionally, public opinion in the country was not in favor of the war as many Turks felt that the war might engulf the whole Middle East and destabilize their country politically and economically.
- There was also the fear that Saddam Hussein would retaliate directly against Turkey if the country were used as a base for invading Iraq.
- At this point, the Turkish parliament voted against stationing United States and coalition troops that might transit the country.

The United States continued to work on an acceptable allied coalition strategy with nations such as Turkey.

---

**Joining Up**

The United States' staunchest ally in forming a coalition for the invasion of Iraq and in support of reconstructing post-war Iraq has been the United Kingdom. Prime Minister Tony Blair carried out shuttle diplomacy to persuade reluctant nations to join the U.S.-led coalition.

In late January 2003, Blair, together with Prime Minister Jose Maria Aznar of Spain, led a drive to persuade other European nations to join the coalition.

- Six additional nations – Italy, Denmark, Portugal, Poland, Hungary, and the Czech Republic – signed on.
- Most glaring was the absence of the two giants of Europe, Germany and France, which both had very strong reservations about military action against Iraq.

(See the site http://www.washingtonpost.com/wp-dyn/articles/A3685-2003Jan30.html for details on coalition building.)

---

*Continued on next page*

---

**Making the Case against Saddam Hussein**

On February 5, 2003, the U.S. Secretary of State, Colin Powell, presented the U.S. case against Iraq to the United Nations general assembly. Using satellite photographs and intercepted audio messages from the Iraqi military, the Secretary of State offered evidence that Iraq

- Still had WMD
- Was not cooperating with the U.N. inspectors
- Had links to Al Qaida (Al Qaeda)

---

**Military Intercepts**

Powell played recordings of telephone messages between high-ranking Iraqi military officers discussing

- "Forbidden ammo"
- Concealment of modified vehicles from U.N. inspectors

---

**Concealment**

In maintaining that Iraq still concealed WMD, the Secretary of State discussed the following:

- Iraq had bulldozed and concealed alleged chemical weapons at the Al Musayyib chemical complex in 2002 and had a series of cargo vehicles and decontamination vehicles moving around at the site.

- He cited intelligence sources showing that the Iraqis had been dispersing rockets armed with alleged biological weapons in western Iraq.

- Iraqi informants claimed that Iraq had 18 trucks that it used as mobile biological weapons labs.

---

*Continued on next page*

**Al Qaida Connection**

Powell also maintained that

> *"Iraq harbors a deadly terrorist network headed by Abu Musab Zarqawi, an associate and collaborator of Osama bin Laden and his al Qaeda lieutenant."*

- Zarqawi spent two months in Baghdad during May and June 2002 getting medical treatment.
- Some members of his group were supposedly based in Baghdad.

**Reaction**

Reaction among the U.S. population to Powell's presentation was generally positive.

- In the United Nations, member states, which had had reservations about military intervention in Iraq, felt that the UNMOVIC should still be given more time to carry out further inspections.
- Because of member states opposition to military intervention and the threat of a veto by other Security Council members, the United States did not request a vote for military intervention in Iraq.

Full text of Colin Powell's U. N. speech is at http://www.state.gov/secretary/rm/2003/17300pf.htm .

# Possible Options for Invading Iraq

**Desert Storm Type Tactics**

This "scenario" would be a replay of the 1991 Persian Gulf War. There would be a more intense air campaign with faster deployment of ground troops than in 1991. The main invasion would come from Kuwait with air strikes from warplanes based in neighboring countries. It would require about 250,000 troops.

- The disadvantage of this option was that it required basing a large presence of Western forces in the area.
- There might have been repercussions from Arab countries reluctant to join the coalition.
- Since it would require a large build-up of troops and supplies, Saddam Hussein might have been able to launch a pre-emptive strike.

**Baghdad First**

This strategy called for taking the capital of Iraq, Baghdad, where the power and communications center of Saddam Hussein's government was located. Surprise was very important here. The aim was a swift strike that would cause a collapse from within. About 100,000 troops were needed.

- A disadvantage was that fighting would primarily occur in the Baghdad region, mostly in urban areas, with high casualties.
- Saddam Hussein's republican guard had been trained in urban warfare and would probably do much of the fighting.

**Rolling War**

Tactics here included large forces seizing and establishing bridgeheads in northern, western, and southern Iraq.

- Dissident groups, deserters, etc., would then rally, pressuring Saddam Hussein's regime internally. Saddam Hussein's government would eventually collapse.
- Kurds in the north and Shia Arabs in the south would play a role in this campaign.
- There probably would not be a direct assault on Baghdad.
- About 300,000 troops would participate in this war.

*Continued on next page*

# Possible Options for Invading Iraq, Continued

**Rolling War** (continued)

- The main disadvantage to this plan was that the Iraqi opposition was splintered; the allies could not be certain that the Kurds and Shias would fully support this plan.
- The Iraqi Government had experience in controlling minorities in these areas.

**Coup or Exile**

Two other options were:

- A surprise coup, sponsored by the CIA, would seize key installations in Baghdad while U.S.-led air forces attacked military targets around Baghdad.
- Some Arab countries tried to persuade Saddam Hussein and his family to go into exile, with promised shelter and safety.

This early interview covering various political, administrative and international angles gives good perspective on the coming war by a PBS correspondent http://www.pbs.org/wgbh/pages/frontline/shows/gunning/interviews/sciolino.html .

(This page intentionally left blank.)

# Operation *Iraqi Freedom*

## Preparation for War

**Selected Topics** The topics covered in this section are shown in the table below.



**Some Basic Assumptions & Conclusions**

United States and coalition forces had two important questions in relation to winning the military campaign in Iraq:

- Could the Iraqis put up a strong resistance?
- Could the United States react as quickly to changes on the battlefield as it thought it could?

Conclusions were:

- Even if the Iraqis could resist effectively and the United States had problems with managing the battlefield, the United States still could win the war.
- Even if there were resistance at first, the U.S. military could isolate southern Iraqi forces.

The United States could then use airpower to take out the northern Iraqi military. It could also isolate and overcome Baghdad without heavy casualties. This combination of mobility and air power would make U.S. success very likely.

*Continued on next page*

**Possible Worst Case**

The United States couldn't guarantee the

- Timing of civilian casualties
- Range of civilian casualties

If intense aerial attacks were necessary, it might take up to two months to be effective. Then a full ground assault would follow in the hottest time, summer.

Civilian casualties would be heavy throughout the country, including Baghdad. Chemical weapons use by Iraq might be a reality in this case too. Even then, the U. S. would probably win the war. The chief issue would be: Is an extended war-fighting scenario politically tenable?

Keep these points in mind as you continue reading: weighing the military and the political is key to understanding the military campaign and later reconstruction of Iraq.

**Military Build-up**

Jan. 7 — The U.S. military built up forces in the Arabian Gulf, but said that it was not a sign that war with Iraq was inevitable (Defense Secretary Donald Rumsfeld).

Jan. 14 — During a White House meeting, President Bush said he was tired of the continuing deception on the part of Iraqi dictator Saddam Hussein and saw no evidence that Hussein was disarming as required by U.N. Security Council Resolution 1441.

Feb. 8 — The Defense Department enlisted commercial airlines to transport troops and equipment as part of the buildup for possible war with Iraq, with the activation of the Civil Reserve Air Fleet.

March 7 —

- U.N. chief weapons inspector Hans Blix reported back to the U.N. Security Council saying Iraqi disarmament would take months.

- The United States and Britain presented a revised draft resolution to the Security Council giving Hussein an ultimatum to disarm by March 17 or face the possibility of war.

*Continued on next page*

**Last Minute Diplomacy**

March 11 —

- The White House rejected any plan to delay by one month the new resolution's March 17 deadline for Iraq to disarm or face military action.

- France and Russia said they would veto any U.N. resolution that implied war.

March 16 —

- The leaders of Spain, Portugal, United States and Britain met in the Azores for a final attempt at diplomacy. This resulted in an announcement that the allies would not pursue a second U.N. resolution.

- U.N. observers evacuated the demilitarized zone along the border between Iraq and Kuwait ahead of the possible war.

**The Ultimatum**

March 17 —

- Bush delivered an ultimatum that Saddam Hussein and his sons have 48 hours to leave Iraq.

- Arms inspectors evacuated the country.

# Operation *Iraqi Freedom:* The War Begins

**"Decapitation" Attack**

March 19 — The war begins. In a "decapitation" attack, a suspected gathering of Iraqi leaders, including Saddam Hussein, in Baghdad was struck by cruise missiles and bombs dropped by F-117 Nighthawk stealth aircraft. The aim was to kill Saddam Hussein in one of his palace bunkers. The attack did not succeed.

**Airpower Focus**

- Coalition air forces also struck Iraqi long-range artillery emplacements, air defense sites and surface-to-surface missile sites.
- Patriot PAC-3 missiles in Kuwait successfully intercepted two Iraqi missiles.
- Coalition aircraft dropped almost 2 million leaflets over Iraq — the second million-plus leaflets dropped in two days. Coalition forces used the fliers and broadcasts into the country to get information to the Iraqi population.

**Air Raids on Baghdad**

Iraqi air defenses in Baghdad were at best sporadic during the coalition bombing of the capital

- Iraqi anti-aircraft artillery (AAA) and surface-to-air missiles at times returning fire against coalition aircraft, and other times remaining silent.
- Additionally, an important focus of the raids was Iraqi command and control centers and communications in the capital.

**Reasons for Poor Iraqi Air Defenses**

There were several possible reasons for the lack of AAA and surface-to-air missiles:

- Coalition bombing was effective in degrading Baghdad's air defense system.
- Coalition forces used B-2 and F-117 fighters, which eluded radar and could launch bombs from high altitudes, making most Iraqi air defenses totally ineffective.
- Iraqi commanders were holding off on using AAA and surface-to-air missiles until they felt the final offensive against the capital was in progress.

*Continued on next page*

**Degradation of Iraqi Air Defenses**

Throughout the early air bombardment phase of the war, coalition forces steadily bombed Iraqi air defenses around the country, showing air supremacy over every aspect of Iraq except Baghdad.

As the air war progressed, witnesses inside Baghdad said that coalition planes began to fly low enough to be seen by those on the ground. This was the point at which coalition forces were certain that they had truly degraded Baghdad's air defense capability.

**Ground War**

On March 20, 2003, the ground war began.

- The 3rd Infantry Division (ID) (Mechanized) rolled out of Kuwait into southern Iraq at 6 a.m. local time and met only slight resistance.

- Special operation forces went into action throughout western and southern Iraq.

- The 1st Marine Expeditionary Force (MEF) and the British crossed into Iraq to seize Iraq's oil fields near Basra.

*Continued on next page*

**Heading for Baghdad**

 By March 22, 2003, the 3rd ID had penetrated almost 100 miles into Iraq on its way towards Baghdad.

 U.S. Marines had captured the port town of Umm Qasr. Umm Qasr, on the Kuwait border about 290 miles southeast of Baghdad, gave U.S. and British forces a port for military and humanitarian supplies and a base from which to clear out any Iraqi resistance in the south.

 Coalition forces launched a massive missile assault on Baghdad. In the next 24 hours, coalition forces struck hundreds of Iraqi targets.

**Map**

The map below should orient you. It has the most important site names mentioned in this section on Operation *Iraqi Freedom.* We will repeat it throughout this "chronology" as you need to have a "geographic" orientation when discussing any military campaign. A detailed chronology of the Iraqi war crisis is at http://www.infoplease.com/spot/iraqtimeline2.html.

The map below comes from:
http://www.lib.utexas.edu/maps/cia03/iraq_sm03.gif

*Continued on next page*

**Navy and Marine Efforts**

- Navy SEALs and coalition special forces seized Iraq's major gas and oil terminals in the northern Persian Gulf and airfields in western Iraq.
- British Royal Marines seized the Fao peninsula.
- The British also focused their efforts in the South at Basra.
- U.S. and Royal Marines captured Umm Qasr.
- The first elements of the 101st Airborne Division (Air Assault) crossed into Iraq.

Most of the chronology of Operation *Iraqi Freedom*, including the war in the North, South and the West is based on reports from the site: http://www.stratfor.com.

# Top Priority: Baghdad

**The Road to Baghdad**

By March 22, 2003 The 2nd Brigade of the 3rd ID had clashed with Iraqi troops, killing 45 of the enemy with no U.S. casualties.

Bypassing urban areas and resistance, the division penetrated 150 miles into Iraq, approximately halfway to Baghdad.

On March 24, 2003, elements of the 3rd ID reached Karbala, within 50 miles of Baghdad. In a furious sandstorm, the 3rd ID battled the Iraqi Republican Guard's Medina Division outside of Karbala.

On March 26, 2003 -

- Two separate columns of Republican Guard vehicles heading south out of Baghdad were struck by coalition air forces.
- Elements of the 3rd ID encircled the town of Najaf, killing approximately 1,000 enemy troops in a three-day battle.
- The 3rd ID and the 101st paused to refit on March 27, 2003. Elements of the 3rd ID fought Iraqi irregulars at the town of Samawah.
- Units of the 3rd ID pushed north to Hillah while the rest of the division advanced to within several miles of Karbala on March 30.

On April 2, 2003 -

- The 3rd ID crossed the Euphrates River at Musayyib and was within 30 miles of Baghdad.
- U.S. Marines closed in on the city from the southeast after crossing a canal and the Tigris River at Numaniyah.

**Into Baghdad**

On April 3, 2003 -

- Part of the 3rd ID fought its way into Baghdad's international airport as another part came to within 10 miles of the city.
- U.S. Marines came to within 15 miles of the city.

*Continued on next page*

**Taking Baghdad**

On April 4, 5 -

- After a tank battle, the 3rd ID took over the entire Baghdad international airport and began encircling the city.
- Marines fought Iraqi soldiers in a fierce fight before advancing to the southeastern edge of Baghdad.
- An armored column of the 3rd ID drove through southwestern Baghdad, encountering pockets of resistance, without any casualties.

April 6 –

- U.S. military planes land at the Baghdad airport.
- The 3rd ID made another sweep through Baghdad, advancing from the west.

**Saddam Hussein's Residences**

On April 7, 2003 —

- Elements of the 3rd ID seized Saddam Hussein's Republican and Sijood presidential palaces.
- U.S. Marines advanced further into eastern Baghdad.

**Collapse of Baghdad: April 9, 2003**

- Soldiers of the 3rd ID continued to search out and destroy remaining pockets of Iraqi resistance in Baghdad.
- U.S. Marines captured the Rashid military air base in eastern Baghdad.

All structured resistance in Baghdad collapsed and the regime's control was broken.

- Baghdad residents came out onto the streets and toppled statues and destroyed pictures of Saddam Hussein and his Baath Party.
- The White House called this an "historic moment," but said the war had not yet ended.
- Saddam Hussein's whereabouts were still unknown.

# War in the South

**On the Road to Baghdad**

As mentioned earlier, the movement to Baghdad came from the south out of Kuwait. This meant that coalition forces had to eventually take the towns from Basra through Nasiriyah and Karbala. Much of this was covered in earlier sections **Operation *Iraqi Freedom*: The War Begins** and in **Top Priority: Baghdad**. In this section you will cover a little more about key regions of the south and their importance.

On March 21, 2003, U.S. Marines captured the strategic port in the southern Iraqi city of Umm Qasr, which is located along the Iraq-Kuwaiti border, about 300 miles southeast of Baghdad. Capturing this town gave U.S. and British forces a port for delivering military and humanitarian supplies, potentially free from Iraqi resistance.

*Continued on next page*

**Basra (Basrah)**

The opening scenario of Operation *Iraqi Freedom* saw Marine expeditionary forces and Army special forces occupying the Basra region because Saddam Hussein's forces had set fire to a number of oil wells in this oil-laden region of Iraq.

- Although Basra is a city smaller than Baghdad, it could have presented a stumbling block to coalition forces advancing on Baghdad. British forces would have rapidly been "depleted" by extensive house-to-house urban combat.
- Any long-term siege of the city, with bombardment, would have meant starving the population, and would have been a true political liability for the coalition.

Basra was important, however, as part of the southern front. It lies on the north-south axis to Baghdad.

# Operation *Iraqi Freedom*: War in the North

**Kirkuk and Mosul: Keys to the North**

Key to the northern front of Operation *Iraqi Freedom* were two important oil-producing centers: Kirkuk and Mosul.

- If coalition forces could occupy these two population centers, the road from the north through Tikrit, Saddam Hussein's hometown region, to Baghdad would be secure. This would allow coalition forces a two-pronged approach, from the north and south, on Baghdad.

- Besides coalition forces, which were much fewer than in the south, the chief player in this northern front were the Kurds and possibly Turkey, which bordered on Kurdish-dominated northern Iraq.

*Continued on next page*

**Initial Phase**

On March 26, 2003, the coalition opened the invasion of Iraq on the northern front. About 1,000 U.S. paratroopers landed in Kurdish-controlled Iraq.

Coalition warplanes bombed:

- Iraqi positions near the town of Chamchamal in Kurdish-controlled northern Iraq. Chamchamal is about 20 miles east of the key oil city of Kirkuk,
- Kirkuk in recent days. Aircraft also were heard flying over the nearby city of Sulaymaniyah.

**Kurdish Involvement**

Early on, Kurdish leaders hesitated in attacking major centers in northern Iraq; the Kurdish commander in the Dohuk area of northern Iraq said that:

- There were too few U.S. troops in the North to open a new front against Iraqi military forces.
- Air raids against Iraqi defensive lines during the early days of the offensive had not been very concentrated or intensive in the North.

**Overrunning Iraqi Positions**

Later it became clear that Iraqi forces in northern Iraq were not a strong match for Kurdish forces. With increased air support, Kurdish, together with U.S. ground forces, made steady progress against Iraqi forces.

A report on Kurdish militia fighting Iraqi forces at: http://www.washingtonpost.com/ac2/wp-dyn?pagename=article&node=&contentId=A44821-2003Mar28&notFound=true

*Continued on next page*

**Mosul**

On April 9, Mosul fell to essentially Kurdish forces supported by U.S. special operations forces and air power.

- The Iraqis had withdrawn some of their divisions from Mosul, weakening its defenses substantially.
- Mosul had limited strategic importance. Kirkuk, to the southwest, was more heavily defended. Kurdish troops would need increased coalition support to take it.

The fall of Mosul did clear the way to Baghdad. The road from the north to Baghdad passes through Tikrit, which is Saddam Hussein's home town and was his political base and supposedly a heavily defended stronghold

**Kirkuk**

On April 10, Kurdish and U.S. forces advanced to the western edge of Iraq's northern oilfields at Kirkuk and moved into the town of Dibis without encountering any Iraqi resistance. The oil facilities around the town appeared to be intact.

Later in the morning the Kurdish forces of Kurdistan Democratic Party and the Patriotic Union of Kurdistan occupied Kirkuk, which Baathist party militants and Iraqi forces had deserted earlier that day.

Report on Kurds taking Kirkuk at http://www.washingtonpost.com/ac2/wp-dyn?pagename=article&node=&contentId=A5395-2003Apr10&notFound=true

*Continued on next page*

**Tikrit**

Because of continued pressure by coalition special operations forces and Kurdish forces in the north, Iraqi forces found it very difficult to withdraw to Tikrit and Baghdad. Tikrit was taken by coalition forces the same way Baghdad and Kirkuk had been captured earlier, with most of Saddam Hussein's forces disappearing before coalition forces had arrived.

- About 20 U.S. tanks entered the center of Tikrit along the main road from Kirkuk and occupied the city square.

- Early on April 14, 2003, a task force of the 1st Marine Expeditionary Force, supported by Cobra helicopters and F-18s, attacked Republican Guard units on the southern edge of Tikrit.

# War in the West

**Purpose**

The western front was a means to control airfields and to suppress any possible threatening missile systems there.

**Special Operations**

U.S. special operations forces gained control of western Iraq, from the Jordanian border, 170 miles to the Mudaysis airfield.

- The area is mostly unpopulated desert but has a few airfields and areas that might have been used by the Iraqi military to hide SCUD missile launchers.
- Coalition special operations forces, including British and Australian commandos, had already captured the airfields at the H2 and H3 pumping stations and were said to have been involved in skirmishes at Ar Rutbah and Akashat.

*Continued on next page*

# War in the West, Continued

**Key Routes**

The area reportedly under coalition control also included two important highways into Iraq:

- The first, from the Jordanian border through Ar Ramadi to Baghdad.
- The second, from Arar, in Saudi Arabia, through Karbala to Baghdad.

## The Turkish "Card"

**A Complex Challenge**

In contrast to the direct attack on Baghdad, the war on the northern front had been more complex in terms of the

- Local population
- Effect on Iraq's neighbor, Turkey

Here coalition forces had to consider a number of variables:

- Chiefly: the Kurds, indigenous to northern Iraq
- Possible military action that Turkey might take in northern Iraq

Between these two "allies" as you saw in an earlier section of this handbook, is much tension, ethnic enmity, and nationalistic sentiment. These problems will play a part of the eventual solution to peace in northern Iraq.

**Turkish Actions**

Turkey took the following actions in relation to its interest in Operation *Iraqi Freedom*:

- Turkey allowed U.S. aircraft to only use Turkish air space, but its parliament banned U.S. military planes deployed in the war against Iraq from taking off from or landing at Turkish bases, including the main air base in Incirlik.

- Some Turkish trucks did assist coalition forces by carrying some supplies for them into northern Iraq.

- Additionally, Turkish soldiers settled into Turkish villages and fields along the border with Iraq, positioning tanks, rocket launchers and artillery behind embankments.

*Continued on next page*

**The U.S. Perspective on U.S - Turkish Relations**

U.S.-Turkish relations probably outweigh many other international considerations in the Middle East. Regardless of Turkish "non-cooperation" during Operation *Iraqi Freedom* the United States needs Turkey.

- Turkey is the key nation in the United States' geopolitical strategy in the region.
- It influences events in the Balkans, the Middle East and in many of the Muslim countries of the former Soviet Union.

**Importance of Turkey**

Turkey is the single most important "piece of real estate" in the Euro-Southwestern Asian area. It has been a basic pivot of U.S. strategy since World War II.

A break with Turkey is inconceivable, regardless of how it treats the United States.

**The Turkish Perspective: Importance of the U.S.**

Turkey depends on the United States to

- Guarantee its national security
- Support Turkish armed forces in guaranteeing Turkish secularism
- Prevent the disintegration of Turkey by controlling the Kurds, who are interested in a unified independent state, which would probably include parts of Iraq, Turkey and Iran

*Continued on next page*

**The Kurds: Forcing a Symbiotic Relationship**

The Kurds have cooperated with the U.S. forces in Operation *Iraqi Freedom* (the Turks did not). For this, the Kurds expect something in return.

- The U.S. administration wants to reward the Kurds, but is limited by the Turks.

- Any move toward a Kurdish sovereign state would destroy U.S.-Turkish relations, and the threat of an independent Kurdish state would be a nightmare to the Turks, who would see it spreading and influencing the Turkish Kurds of southeastern Turkey.

As you can see the Kurds are one of chief factors that keeps Turkey and the United States balanced in a dependent relationship.

**The Turkoman Minority**

To make the Turkish-Kurdish issue even more thornier, the Turkoman, a large Turkic minority living in Kurdish-controlled northern Iraq, are beginning to demand power in that region.

- There have been clashes between Kurds and Turkomans in the Kirkuk area.

- Naturally, Turks favor their "brothers" the Turkoman in any kind of "struggle" with the Kurds.

As you can see, ethnicity will continue to play a dominant role in any political stabilization of post-Saddam Hussein Iraq.

# Statistics: Operation *Iraqi Freedom*

**Length**

Combat operations lasted officially March 20 – May 1, 2003

**Official Names**

- U.S. – Operation *Iraqi Freedom*
- United Kingdom - Operation *Telic*.
- Australia - Operation *Falconer*

**Troops Deployed**

More than 300,000 coalition troops were deployed to the Gulf region:

- 255,000 U.S.
- 45,000 British
- 2,000 Australian
- 200 Polish

**U.S. Estimates**

- Casualties – 115 U.S.; 53 other coalition countries
- Wounded – 2,110
- POWs – 8 (all rescued)

**Iraqi Estimates**

- Iraqi troops – 350,000
- Combat deaths – 2,320
- Iraqi civilian deaths – 7,900-9,600
- Iraqi civilian wounded – 5,100
- Iraqi POWs – 7,300

**Cost**

$4 billion per month to keep troops stationed in Iraq

Statistics from http://www.infoplease.com/ipa/A0908900.html

**(This page intentionally left blank.)**

# After Operation *Iraqi Freedom*

## Important Political Events in Post-Saddam Hussein Iraq

**Selected Topics** The topics covered in this section are shown in the table below.



**Administrative Structure**

A civilian administrator, L. Paul Bremer, former ambassador and State Department counter-terrorism director, currently runs the U.S. administration of post-Saddam-Hussein Iraq. . (He replaced retired Lieutenant General Jay Garner, the first administrator.)

An excellent time line of post-Saddam Hussein administrative and managerial events is at http://www.cfr.org/reg_index.php?id=6l35ll12.

**The Most Wanted List**

In April 2003, the United States developed a list of the 55 most-wanted members of Saddam Hussein's regime in the form of a deck of cards.

- One of the earliest "members of this deck" taken into custody was Saddam Hussein's former deputy prime minister Tariq Aziz.
- In May 2003, Saddam Hussein's sons Uday and Qusay Hussein were killed in a gun battle at a villa in Mosul.
- In August 2003, Saddam Hussein's cousin, Ali Hassan al-Majid or "Chemical Ali," was captured.
- In December 2003, in a night operation called Operation *Red Dawn*, U.S. forces captured Saddam Hussein in a village near Tikrit. He was hiding in an underground "spider hole" with a few weapons and $750k.

Rewards have been offered for information leading to the capture of the remaining "cards."

*Continued on next page*

# Important Political Events in Post-Saddam Hussein Iraq,
Continued

**Early Administrative Measures**

In May 2003, the UN Security Council approved a resolution

- Supporting the U.S-led administration in Iraq
- Lifting of economic sanctions

The U.S. administrator in Iraq abolished

- The Baath Party
- The Iraqi army and police force as constituted under Saddam Hussein
- Institutions of Saddam Hussein's regime

The interim governing council (IGC) of Iraq met for first time.

**Contracting for Rebuilding Iraq**

Two large U.S. corporations are participating in the reconstruction of Iraq. Halliburton Corporation was awarded a contract by the U.S. Agency for International Development for more than 1.7 billion dollars. A Halliburton subsidiary, Kellogg Brown and Root, was awarded a contract to

- Operate Iraqi oil fields
- Distribute oil from Iraqi fields

Bechtel was also awarded a contract to rebuild bomb-damaged Iraq.

# "New" Iraq, Old Neighbors

**Syrian Challenge**

A democratized Iraq could eventually present a potential threat to many Middle Eastern countries. An example is Syria, which could pose destabilizing problems for both coalition forces and the newly emerging post-Saddam Hussein Iraq.

- Syria voted in favor of Security Council Resolution 1441 and maintained a behind-the-scenes dialogue with the United States despite condemning U.S. war plans.
- Then it changed course, encouraging anti-war activity and sentiment. The president of Syria spoke out about what his country considers the U.S. potential for dominating the Middle East.
- Initially, the country left its border with Iraq open so that anti-coalition voluntary forces could freely mass in Syria and launch "attacks" against coalition forces. Busloads of anti-coalition volunteers launched from Syria into Iraq.
- Later, because of U.S. pressure and threats, the government closed the border though traffic in supplies to anti-coalition forces in Iraq continued.

**Iran**

As you learned earlier, Iraq and Iran have been at odds over the years in their efforts to dominate the Middle East. Additionally, the "favored" status of Iraq's Shias by the regime in Tehran has been a sore spot in the relationship between the two countries.

- It is possible that the pro-Iranian Shias of Iraq could organize to carry out a guerilla style war against coalition forces. Iran has tremendous resources at its disposal to support this kind of a war, something that is uppermost in U.S. policymakers' minds.
- Iran is at best unpredictable in its attitude and actions regarding the liberation and reconstruction of Iraq. While Syrian actions can be contained and are often very evident, Iran has more long-term goals, which in their complexity and subtlety, are difficult to assess.

# New Iraqi Government

**What Iraq needs: Peaceful Pluralism**

Political and economic stability in post-Saddam Hussein depends on a government of all peoples of present-day Iraq.

- The transition to political pluralism is not easy because Iraqis had lived more than three decades under Saddam Hussein's one-party dictatorial system.

- A peaceful means for political parties to coexist, with respect to their differences, must be ensured. This political pluralism, which is one of the most elementary foundations of democracy, is a process that requires time to shape.

---

**The Transfer Process**

The aim of the transfer of power from the U.S. and coalition administration to the newly formed Iraqi government is a progressive shift of power, together with the consolidation of new Iraqi political institutions. Such a process requires time.

---

**Establishing the IGC**

The 25-member Interim Governing Council of Iraq was established in July 2003, based on negotiations between the

- Coalition Provisional Authority
- Former Iraqi anti-Saddam Hussein opposition parties

---

**IGC Responsibilities**

The council:

- Appoints and dismisses ministers.
- Controls the national budget.
- Oversees the convening of a conference to draft a new Iraqi constitution.

General elections are set to follow a referendum on this new constitution.

---

*Continued on next page*

**IGC Make-up**

Most of those holding council seats are

- Members-in-exile from historically anti-Saddam Hussein groups.
- Kurds from the liberated Kurdish areas of Iraq.
- Representatives of indigenous political parties some of which are at odds with the United States.

**UN Approval**

In August 2003, the U.N. passed a resolution

- Welcoming the IGC of Iraq to the U.N.
- Giving a 12-month mandate to the new U.N. assistance mission for Iraq

**Council Action**

- In early September 2003, the IGC appointed a cabinet of 25 ministers, reflecting Iraq's ethnic and religious composition.
- The cabinet was tasked to run the country's affairs until a new elected government was established.

**Reaction to the IGC**

Many Arab countries have not yet recognized the IGC as the government of post-Saddam Hussein Iraq because they consider the ICG a "puppet" of the United States.

France and other countries, which had been originally opposed to the invasion of Iraq, would like to see a more speedy transition of power to the Iraqi people.

## Guerrilla-Style War

**Low Intensity**

CENTCOM has admitted that in its efforts to stabilize Iraq and to maintain the peace there, coalition forces in Iraq face a terrorist adversary using guerrilla style low intensity tactics.

**Types of Opposition**

The terrorist opposition is probably made up of the following types of individuals and groups or has attracted similar individuals

- Remnants of military forces still loyal to Saddam Hussein
- Remnants of the Baath Party
- Sunni Muslims loyal to Saddam Hussein
- Disgruntled Iraqi citizens who have lost patience with the U. S.-led coalition
- Al Qaida (Al Qaeda) and other voluntary terrorist-type forces who have crossed over into Iraq to "liberate" the country from coalition forces

**Terrorist Incidents: Examples**

- In August 2003, the Jordanian Embassy in Baghdad was bombed, with 11 people killed. Twenty-two people were also killed when a terrorist bomb destroyed U.N. headquarters in Baghdad. A car bomb killed 125 people in Najaf, including the prominent Shia leader Ayatollah Mohammed Baqr-al-Hakim.

- A female member of the Interim Governing Council was assassinated in September 2003. Dozens of people were killed when the Red Cross office in Baghdad was bombed.

*Continued on next page*

## Guerrilla-Style War, Continued

**Shift in Tactics?**

- There may be a change in the targets of terrorist attacks. Previously most of the targets had been U.S. personnel and officials associated directly with U.S. reconstruction efforts. Since late October 2003, more Iraqi citizens appear to be targeted.

- The brutal suicide bombing of an Italian police base in Nasiriyah in November 2003 may also signal that terrorist forces have shifted their attacks to regions located outside the capital.

- The murder of Japanese diplomats and Spanish civilians working in Iraq seemed to indicate that terrorists are targeting allied personnel in an effort to discourage coalition countries from sending support to U. S. forces in Iraq.

This discussion is based on http://news.bbc.co.uk/1/hi/world/middle_east/737483.stm.

# Operation *Iron Hammer*

**Operation *Iron Hammer*: A Response**

To deal with the upsurge in terrorist activity (examples in the next section), coalition forces initiated Operation *Iron Hammer* in November 2003.

- This operation was an aggressive attack on insurgents before they could strike coalition forces and coalition-friendly indigenous groups.
- The coalition said it would use "*overwhelming force*" to suppress these attacks.

**Terrorist Activities**

- Insurgents fired on a U.S. supply convoy north of Samara; U. S. forces returned fire.
- Gunmen assassinated Hmud Kadhim, the Education Ministry's director general in Diwaniyah province in the southern town of Diwaniyah. This kind of assassination is part of the new terrorist policy to murder Iraqis who are working with coalition forces.
- Two policemen were wounded when assailants tossed a grenade at a police station in Mosul.
- A roadside bomb exploded in the southern city of Basra as a British civilian convoy was passing by, damaging a vehicle

**Coalition Responses**

- A pair of 2,000-pound satellite-guided bombs was dropped near Baqouba, 30 miles northeast of Baghdad, on camps suspected of making bombs.
- Near Kirkuk, fighter-bombers dropped 1,000-pound bombs on terrorist targets.
- U.S. forces targeted an abandoned dye factory in southern Baghdad that was hit by artillery and air strikes.
- U.S. forces also used heavy artillery, battle tanks, attack helicopters, F-16 fighter-bombers and AC-130 gun ships to take out targets in central Iraq, including Tikrit, Baqouba and Fallujah.

Details can be found at http://www.ctnow.com/news/custom/newsat3/sns-ap-iraq,0,7589745.story?coll=hc-headlines-newsat3

*Continued on next page*

**Battle of Samara**

This was the largest post-Saddam Hussein engagement for coalition forces. In late November 2003, coalition forces engaged insurgent forces, in a planned attack, in which scores of Fedayeen-like troops were routed or destroyed.

**Capture of Saddam Hussein: Implications**

With the capture of Saddam Hussein in December 2003, a number of key issues came to the forefront:

- Would his capture decrease or increase the insurgent attacks on coalition forces and Iraqis working with coalition forces? The chief issue here was how involved in the insurgency "movement" was Saddam Hussein. From the way he was captured, hiding alone in an underground hole, coalition forces cannot be certain how much direction or command he had over such forces.

- An important issue that has been a constant problem for the Bush Administration was Iraq's weapons of mass destruction. Hopefully, Saddam Hussein could reveal his plans and the locations of these weapons and missile systems.

- If Saddam Hussein were, in fact, a key player in the insurgent terrorist attacks on coalition forces and Iraqis, he would be able to give critical information on the structure, location, strength, and leadership of these groups.

# Reconstructing Iraq: Nation Building

**Utilities and Sanitation**

Although the water treatment system was antiquated and damage to the city power system also affected water treatment, about 60% of Baghdad now receives adequate water. The sewage system was not designed to support the 5.6 million inhabitants of metro-Baghdad. Because of looting and sabotage, untreated sewage is now deposited into the Tigris River.

- Baghdad has electricity about12 hours per day.
- The water authority has restored the worst parts of the system.
- U.S. Agency for International Development is in charge of the contract for equipment for water treatment.
- The coalition has allocated more than $100 million for repairs to the sewage system.

Based on:
http://graphics7.nytimes.com/images/2003/11/03/international/1104HEAL.gif

**Health**

A health assessment of hospitals found that 65% of the equipment in the country was not working. In 2003, coalition allies are spending $422 million on health care. All 240 hospitals and 1,200 primary care clinics are now open.

**Education**

Since 1991, there had been little maintenance on school buildings. University and higher technical schools had been isolated from the international academic community. Improvements in education include the following:

- School registration at the primary and secondary level has increased.
- Teachers' salaries have increased.
- Grants, in conjunction with U.S. universities, have been awarded to Iraqi centers of higher education to encourage study in the fields of archaeology, environmental research, agriculture, and legal education.

*Continued on next page*

**Price of Reconstruction**

- About five U.S. soldiers have been killed per week since major combat ended.
- The bounty paid to Iraqis who kill U.S. soldiers is $300.00 – 1,000.00.
- Coalition allies have destroyed about 100 weapons and munitions per day.
- The average price of a Kalashnikov rifle is now about $80.00. In May 2003, it was about $5.00.

**Reconstruction: Eyewitness Report**

This section is based on the article below from http://www.opendemocracy.net/debates/article-2-95-1573.jsp#. It is a first-hand account of some of the every-day "realities" of reconstruction in Baghdad.

**Eight days in Iraq**
by Yahia Said
6 - 11 - 2003

Key points made by the author, who spent eight days in present-day Iraq found "a people yearning for freedom, normality – an end to violence." He had been out of country for twenty-five years.

- A "rediscovered humanity" – the courtesy and politeness of the local Iraqi police force would make inhabitants of Iraq's neighboring countries jealous: " 'I apologise for the inconvenience,' said the Iraqi policeman." The police force is the first government service organization in 35 years to work outside of Baathist control. Said feels this is a plus for the Iraqi people and to the detriment of Saddam Hussein.
- "The wave of lawlessness which gripped the country upon the fall of the regime has largely subsided. Night curfews have been lifted and Baghdad residents are gradually venturing out of their homes. Many now eat out in outdoor cafes and restaurants, instead of hurriedly grabbing takeaways. Goldsmiths on Karrada Street, one of the few revived shopping destinations in the capital, displays its wares late into the night, a big change apparently from just weeks ago."

*Continued on next page*

**Reconstruction: Eyewitness Report** (continued)

- Baghdad is alive with commercial activity. Government employees are now receiving salaries ten times higher than before the war. The streets are clean. Children from poor families Said visited were eating bananas – something unheard of, under Saddam Hussein.

- Many Iraqis forced into exile by Saddam Hussein's regime are now returning to the city to set up political parties, start newspapers and to begin business enterprises.

- Most dangerous is the ongoing political violence, according to Said. Most suicide bombers and terrorists target police stations, schools and international organizations. Many of these terrorists are considered foreigners.

- "Some reports from the country seem to imply that Iraqis welcome at least some of the violence as legitimate resistance to occupation. During my eight days I did not meet a single person who shared this view. Indeed, the vast majority of people in Iraq – especially women, who represent 60% of the adult population – do not want the Americans to leave anytime soon.

- "Iraqis want to regain control of their future and to become truly independent and free. They are sick of violence of all kinds, whatever the pretexts. They are tired of empty rhetoric and ideologies. Most Iraqis want to pick up the pieces of their shattered lives and move on. Whoever cares for them should help them do just that."

# References

**Sources**

Besides the numerous web sites cited in this handbook, the list below of other sources may also help you:

Henderson, Simon. *Instant Empire: Saddam Hussein's Ambition for Iraq.* San Francisco: Mercury House, 1991. Henderson uses a balanced approach to discussing Saddam Hussein and his importance to the Arab and non-Arab world. He covers important subjects such as the family and political background of Saddam Hussein's rise; an analysis of opposition and corruption in Iraqi politics, and the geopolitical motives behind Saddam Hussein's invasion of Kuwait.

Karsh, Efraim and Inari Rautsi. *Saddam Hussein: A political Biography.* New York: Free Press, 1991. Karsh and Rautsi have written an analytical political biography of Saddam Hussein. The book "traces the meteoric transformation of an ardent nationalist and obscure Ba'th party member into an absolute dictator." The authors cover the personality of Saddam Hussein and how it affects his political actions.

Makariya, Kanan. *The Republic of Fear.* Berkeley: University of California Press, 1998. Makariya discusses Iraq as a one party state, the total control exercised first by the Baath Party and then by Saddam Hussein. The emphasis is on the security and secret forces of Iraq and the way they terrorize citizens.

Mylroie, Laurie. *Study of Revenge: The First World Trade Center Attack and Saddam Hussein's War against America.* Washington, DC: AEI Press, 2001. This is an extended analysis of the origins, background, planning, and eventual first bombing of the World Trade Center. The authors focus on the link between the bombers and Saddam Hussein.

Pollack, Kenneth. *The Threatening Storm: The Case for Invading Iraq* New York: Random House, 2002. The author, one of the leading national security advisors during the Clinton administration, concludes that the post-Gulf War policy of containment against Iraq is no longer effective and believes that Saddam Hussein must be deposed by invasion.

*Continued on next page*

**Sources, (continued)**

Sciolino, Elaine. *The Outlaw State: Saddam Hussein's Quest for Power and the Gulf Crisis*. New York: Wiley, 1991.

Trevan, Tim. *Saddam's Secrets: The Hunt for Iraq's Hidden Weapons.* London: Harper Collins, 1999. This book, by one of the leading United Nations weapons inspectors, focuses on how United Nations weapons inspectors uncovered Iraq's secret nuclear, chemical, and biological weapons programs and exposed the deceit that is at the root of Saddam Hussein's regime.

The site below has a chronology of the UN resolutions on disarmament inspections and Saddam Hussein's obstructionism. http://www.cia.gov/cia/publications/iraq_wmd/Iraq_Oct_2002.htm#03.

The following is a list of online "bibliographical" sources that have extensive sites dealing with Iraq and Operation *Iraqi Freedom.*

http://www.au.af.mil/au/aul/bibs/iraq/iraq_crisis.htm

http://www.au.af.mil/au/aul/bibs/tertod/tertod.htm

http://www.centcom.mil/operations/iraqi_freedom/iraqifreedom.asp

http://www.lib.mich.edu/govdocs/usterror.html

http://fas.org/man/dod-101/ops/index.html

http://news.bbc.co.uk/1/hi/world/americas/1584660.stm

http://www.ringnebula.com/Oil/Timeline.htm

http://www.cfr.org/reg_index.php?id=6|3|5||1

http://www.army.mil/

# Appendix A
# Islam

(This page intentionally left blank.)

# slam: Basic Ideas

**ntroduction**

You have probably heard the name *Islam* associated with Iraq. Most people in Iraq and in surrounding countries believe in Islam. In this section, you will learn about Islam, its main ideas, customs, and practices.

(http://thetruereligion.org/intro.htm - - a short introduction to Islam)

**)efinition of slam**

Islam means submission to the will of God in the Arabic language. The word *Islam* has the same root as the Arabic word *salaam*, which means peace. Muslims around the world often greet each other with the phrase "Salaam" in much the same manner as Jews say, "Shalom," as a greeting. Both salaam and shalom come from the same common language root, the three-letter root - - *s-l-m.*

(http://media.isnet.org/off/Islam/basics/index.html) – discussion of various "doctrinal" aspects of Islam by noted theologians.

**\llah**

Islam is not named after a person like Christianity, which was named after Jesus Christ, Buddhism after Buddha, and Confucianism after Confucius. The central focus of Islam is always God.

The basic tenet of Islam is that you must submit to Allah (the Arabic word for God) and live according to His divinely inspired law.

**The Precept**

- The key truth that Allah has revealed to mankind is that the only divine and worshipful "being" is Allah, the almighty God; thus all human beings should submit to and worship Allah.

- Allah has 99 names, among them, The Gracious, The Merciful, The Beneficent, The Creator, The All-Knowing, The All-Wise, The Lord of the Universe, The First, The Last, and others.

No matter what "sect" of Islam a person belongs to, he believes in this important religious belief that is the common thread of Islam.

# Arabs and Islam

**Muslims (Moslems)**

Now that you know that the religion is called Islam and the name of the God of Islam is Allah, what does the word *Muslim* mean? The word has the same s-l-m root as the words *Islam* and *salaam*. A Muslim is a person who submits to the will of Allah, regardless of race, nationality, or ethnic background. A Muslim has to submit completely and obediently to Allah and live according to Allah's message.

- Muslims worship Allah alone and must not worship any person, place or thing other than Allah. They believe that Allah is the God for the Christians, the Jews, the Muslims, the Buddhists, the Hindus, and even atheists.
- Orally repeating the basic belief of a Muslim found in the "motto" of Islam: "*Laa Elaaha illallaah*" which means, "There is no god but Allah," is the way a Muslim professes that he belongs to Islam.

**Importance of Arabic in Islam**

- The "motto" above, "*Laa Elaaha illallaah*," is in Arabic.
- The prayers, sayings, and many blessings of Islam are all spoken in Arabic, just as the holy book of Islam, the Qur'an (Koran) is written in Arabic.
- This common religious language of Islam is one of the ties that bind all Muslims together.

*Continued on next page*

**Arabs and Non-Arab Muslims**

Because of the importance of the Arabic language in Islam, many people assume that most Muslims are Arabs. This is not the case: About 80% of all Muslims are **not** Arabs.

- In fact there are more Muslims in Indonesia than in the Arab Middle East.
- Besides Indonesia, Muslims make up the majority in such non-Arabic countries as Turkey, Pakistan, Uzbekistan and other central Asian countries of the former Soviet Union.
- There is a large minority in China too.
- In Europe, Albania is Muslim. Bulgaria, Bosnia, and Macedonia have large Muslim populations. Elsewhere in Europe, there are immigrant communities of Muslims from Africa, Turkey, and Asia in France, Britain, and Germany.
- In the Americas Muslims have increased in recent years, both from conversions and immigration; 20% of the population of Suriname in South America is Muslim. Today there are about five million Muslims in America.

# Muhammad (Mohammed)

**Importance**

*Muhammad is … the Messenger of God*
*and the last of the prophets …* (Qur'an, 33:40)

This one phrase from the Qur'an shows Muhammad's place in Islam and in world religion. Muslims believe that Muhammad is the prophet of Islam. His message is Islam. The revelation of Islam that Muhammad received is the *Qur'an* (Koran).

**Muhammad and Other Religions**

In terms of other world religions, Muhammad is the last prophet of God to mankind.

- He is the last messenger of God.
- His message applies to all mankind, regardless of their beliefs.
- Muhammad is the successor to Moses, Jacob, Isaac, Abraham, and Jesus.

Muhammad is **the** prophet among all previous prophets and messengers.

- He reinterpreted and corrected the message of Allah and gave mankind the true word from Allah. In a sense, Muslims consider that he "cleansed" the message.

- In spite of his importance in Islam and for the world, Muslims do not worship him or ascribe any divine qualities to the prophet.

**Muhammad, His Calling**

Muhammad was born in Mecca (Makkah) in what is now Saudi Arabia, in the year 570 CE. His father died before he was born. Soon after his birth, his mother also died, so Muhammad was raised by his uncle.

Because of his trustworthiness and sincerity he was often asked to arbitrate disputes. He has been described as calm and meditative. Muhammad was very religious. It was his habit to meditate in a cave near Mecca.

At the age of 40, during one of his retreats, Muhammad received his first revelation from Allah through the Angel Gabriel. This revelation, which he continued to receive for twenty-three years, is known as the *Qur'an* (Koran).

*Continued on next page*

**Spreading Islam**

As Muhammad began to preach the truth that Allah had revealed to him, he and his small group of followers suffered persecution.

- This became so strong that in the year 622 CE Muhammad and his followers emigrated from Mecca.
- This event, called the *Hijra* ('migration'), when they traveled about 260 miles to the north to the city of Medina, marks the beginning of the Muslim calendar.
- After several years, Muhammad and his followers returned to Mecca, where they made peace with their enemies and established Islam.
- Before Muhammad died at the age of 63, most of Arabia was Muslim, and within a century of his death, Islam had spread to Spain in the West and as far east as China.

# Five Pillars of Islam

**Introduction**

The five tenets are the foundation for all Islamic belief. They are the cornerstones of Islam. ( http://www.usc.edu/dept/MSA/fundamentals/pillars/ – an introduction to the five pillars.)

**Commitment**

Called the Ash-Shahaadah, this is the "credo" of Islam:

*"I bear witness that there is none worthy of worship except Allah and I bear witness that Muhammad is His servant and messenger."*

All Muslims subscribe to and repeat this belief.

**Prayer**

Called As-Salaat in Arabic, this is the requirement that adult Muslims pray at five specific times during a twenty-four hour day to Allah.

- This is a ritual prayer requiring Muslims to say specific prayers and to coordinate hand and arm movements, eventually culminating in complete prostration or bowing down with the hands on the floor to show complete submission to Allah.

- While performing As-Salaat, Muslims face in the direction of Mecca.

**Alms**

Known as the Az-Zakaat in Arabic, this means that Muslims must contribute to the support of those less fortunate Muslims--support of the poor. Muslims are encouraged to pay their Zakaat during the holy month of Ramadan.

*Continued on next page*

---

**Fasting**

In Arabic this is called As-Sawam. It means that during the month of Ramadan, adult Muslims who are in good health

- Cannot eat, drink, or smoke from sunrise until sunset every day of this holy month.

- Must abstain from sexual relations.

The intention of fasting is to worship Allah. Ramadan is a sacred month because it commemorates the time when Allah first revealed the Qur'an to Muhammad. The time for Ramadan changes each year because it is based on a lunar calendar. Ramadan comes 11 days earlier each year.

---

**Pilgrimage**

Called Al- Hajj (Hadsh), this pilgrimage requires a Muslim in good health to travel to Mecca in Saudi Arabia at least once in a lifetime.

- Specific preparation is necessary for the Hajj; for example there are rules about shaving and cutting hair and what to wear.

- One of the high points of the Hajj is the *umrah*, which means performing a ritual walk seven times around the sacred stone of Islam called the Kab'ah.

- The Hajj commemorates the sacrifices and faith of Abraham, his second wife, Hagar, and their son, Ishmael.

According to the Council on Islamic Education, it is the largest, regularly scheduled international gathering on Earth.

---

# *Al Qur'an* (Koran) and Sunnah

**Definition**

Al Qur'an (the Qur'an) is the revealed word of Islam, the holy book of the Muslims.

- Muhammad received the message in bits and pieces from the Angel Gabriel during a period of 23 years.
- The word *Al Qur'an* means the recitation. During Muhammad's lifetime it was recited publicly.
- The original language of the Qur'an is Arabic.

It is important to remember that Muslims believe the Qur'an is **the** word of Allah, untouched and uninterpreted by human beings. As Allah's final message to mankind, it is the holy book that supercedes all others: the *Old Testament*, the *Gospels*, etc.

**Form and Content**

The Qur'an is a collection of surahs or chapters, many of which you might call verses.

- Muhammad received the surahs from Allah. Scribes selected by Muhammad usually wrote down these verses. Sometimes they wrote on wood, trees, parchment, and even stones. Many followers also memorized the Qur'an by heart.

- In later years, the Qur'an was recopied and refined; accents and markings for reading were added.

**Interpretation**

- The Sunnah is the spoken word and acts of Muhammad. The Sunnah, Muhammad's actual words and deeds, uses the life of Muhammad to explain and expand on the verses and teachings of the Qur'an so that Islam can eventually become a world religion.

- A second form of interpretation is called hadith. This is narration about the life of the prophet or what he approved - as opposed to his life itself, which is the Sunnah.

The Qur'an, the Sunnah, and the ahadith (plural of hadith) are the basic body of Islamic religious doctrine for all Muslims. (http://www.afghan-network.net/Islam/ ) (http://www.usc.edu/dept/MSA/)

# Jihad (Djihad)

**Definition**

The word *jihad* means struggle in Arabic. This means a struggle between the forces of Allah (good) and the forces of evil --an eternal struggle.

You have probably heard the word *jihad* in reference to Osama bin Laden and the Taliban. They consider the terrorism that they perpetuate a jihad. Actually, the word *jihad* has a couple of interpretations.

**Self-defense**

Islam allows fighting in self-defense and in defense of religion. There are rules of combat including prohibitions against harming civilians and against destroying crops, trees and livestock.

If good Muslims do not risk their lives in the cause of Islam, injustice would triumph. The Qur'an says:

> " *Fight in the cause of God against those who fight you, but do not transgress limits. God does not love transgressors.*" (2.190)

War is the last resort and is subject to the precepts of Islamic sacred law.

**Inner Struggle**

The second and more personal meaning of jihad is an inner struggle. Each of us, according to Islam, is constantly fighting a "war" against self-centered desire, and egotism, with the final goal of attaining inner peace.

# Islamic Schools (Sects)

---

**Religious vice Secular Law: Western View**

The United States and most other Western countries make a clear distinction between the rule of law and religion.

- Religion is a separate personal issue that does **not** determine law.
- A legal case may make use of religion, but religion is not the determining factor in most legal cases.
- Simply put, a religious denomination, such as Methodism, Catholicism, or Judaism, doesn't have the power to rule on a case of adultery in a civil court or on a case of fraud in a criminal court.

The United States has a secular system of law; religion plays no direct part in such a system.

---

**Islamic View**

If you look at Islam, the perspective is different. Islamic law consists of guidelines and rules that determine all aspects of a Muslim's life from how to perform ritual prayer to conducting business transactions.

- It also includes crimes and the appropriate punishments for each.
- These Islamic laws are based primarily on the Qur'an and are called the Sharia.
- Muslims must apply these precepts handed down to Muhammad in the sixth century to present-day situations.

---

**Interpretation and Re-interpretation**

This kind of religious legal system, particularly how to interpret the Qur'an and the ahadith, has resulted in a certain degree of divisiveness in Islam, just as varied interpretations of Christian doctrine have caused Christianity to split into various denominations. The two main sects of Islam, which are the schools of Islam, are the

- Sunni
- Shi'ites (Shi'ia)

---

*Continued on next page*

**Shi'ites (Shi'ia)**

The basic divisiveness in Islam is based on an historical incident:

- After Muhammad died, there was confusion about who should succeed him as the leader of Islam. One group felt that the successor was Imam Ali who had been appointed by Muhammad according to Allah's decree. Ali was a relative of Muhammad and the first to accept Islam.

- This group, mostly from the household of the prophet, was a minority and became known as the Shi'ites. They also believe that the leader of Islam must be endowed with grace and benevolence and should be infallible.

**Sunnis**

The majority group, called Sunni Muslims, believe that:

- Muhammad did not choose a specific successor and probably assumed that after his death, Muslims would find their own leader.

- The prophet did not tell his followers how they ought to select their leaders or what qualifications their leaders should have.

The Sunnis chose a leader, later called the caliph, from outside Muhammad's household. The Shi'ites refused to accept the Sunnis' choice and split off from this main group.

**Results of the Division**

Because of this succession split, variations in Islamic doctrine, law, and practice have developed between these two main groups over the centuries.

- Sunni Muslims predominate in Saudi Arabia, Afghanistan, Pakistan, Turkey, and Indonesia.

- Shi'ites are the majority in Iran and southern Iraq.

Sunnis make up about 83 percent of Muslims, according to the *Encyclopedia Britannica*; Shi'ites, about 16 percent; and a few other small groups, the remainder.

# Osama (Usama) bin Laden and Islam

**Bin Laden's Religious Background**

Bin Laden's brand of Islam is based on extremely conservative Islam called Wahhabism, which was espoused by Muhammad bin Abd al-Wahhab (1703-87). He was the founder of the sect and the co-founder of Saudi Arabia. The Wahhabis and the royal Saud family have long intermarried. The Saudis have adopted many of the conservative precepts of Wahhabism.

**Wahhabism: The Way of a Muslim Life**

Wahhabi Islam is a total religious system. It has an answer for every question about a Muslim's life including social, legal and spiritual aspects. Wahhabi Islam is ascetic.

- Men should wear short robes and even avoid the black cords used on turbans and headgear.
- Mosques should have no decoration.
- Drinking alcohol is forbidden.

Punishment is based exclusively on the Qur'an, for example:

- The right hand should be amputated for theft.
- Adulterers should be stoned to death.
- Murder and sexual deviation are punishable by beheading. To this day Saudi Arabia uses these punishments, especially beheading for capital crimes.

**Bin Laden and the Taliban**

The kind of radically, conservative state that the Taliban had developed in Afghanistan had roots in Wahhabism, which inspires and feeds on the Taliban's fundamentalist religiosity.

- For Islamic fundamentalists, the Taliban have created what approaches a pure, ideal society based on Islamic religion and law.
- This society fits bin Laden's mold too.

*Continued on next page*

**Focus on Expansion and Terrorism**

- While bin Laden follows the conservative Wahhabi tenets, what he most fervently supports is that his brand of Islamic faith must expand. (To support this kind of expansion, Saudi Arabia supported the Afghan *mujahidin* when they were fighting the Communists. So did Osama bin Laden.)
  http://www.newyorker.com/FROM_THE_ARCHIVE/ARCHIVES/?010924fr_archive03 – good article on bin Laden, his life and religion

- Now bin Laden regards the struggle as a war between two civilizations: His Islam and the non-Muslim civilization, specifically the United States. Toward winning this struggle, bin Laden has committed his forces to terrorism.

http://www.brookings.org/fp/projects/terrorism/faqs.htm#qb4 – a set of informative questions about terrorism.

**Bin Laden-sponsored Terrorism**

- Besides the infamous 9/11 terrorist attacks against the Pentagon and the twin towers, bin Laden allegedly sponsored the attacks on American embassies in Africa.

- U.S. and Russian officials are also concerned that he is financing Chechen rebel operations out of Dagestan, a former Soviet republic in Asia.

A primary concern is that bin Laden's forces are attempting to acquire or build chemical and/or biological weapons to use in their terrorist attacks. In the case of the Chechen rebels, he may be trying to help them acquire radiological dispersal devices.

*Continued on next page*

**Al Qaida (Al Qaeda)**

*Al Qaida* (Al Qaeda) is the name of the network of Islamic extremists that bin Laden has at his command to carry out his radical Islamic terrorism.

- Al Qaida consists of a group of about 3,000 commanders. Troops of Afghanis and Pakistanis number 200,000+.
- There are a number of terrorist centers, with numerous cells, in North America, Yemen, Saudi Arabia, Albania, Kosovo, Algeria, Chechenya, all the former Soviet republics of Central Asia, the Philippines, Egypt, Ethiopia and Somalia.

**Bin Laden's "Army"**

Al Qaida is bin Laden's personal "army." Bin Laden could probably rally an army of over 100,000+ men around the world, excluding those currently "in service" in Pakistan and Afghanistan.

- Many members of his overseas forces are ready for both sustained and one-time operations.
- In the interim, the members of these terrorist forces return to normal "civilian" life in host countries. They are also available for terrorist operations on their own countries.

**Funding for Al Qaida (Al Qaeda) Terrorism**

The primary sources for funding Al Qaida-sponsored terrorism have been:

- The sale of drugs, primarily opium.
- Contributions from Saudi Arabia through various banks and from various Saudi businessmen.
- What might be called tribute money to Al Qaida from countries that do not want the organization to establish cells in their own countries.
- Funds from bin Laden's personal fortune.

*Continued on next page*

**Bin Laden and Saudi Arabia**

Even though the country where he grew up, Saudi Arabia, is extremely conservative, bin Laden finds that it is not conservative enough. He thinks the following reasons, which are contrary to his purist concept of Islam, are grounds for overthrowing the Saudi royal family:

- Alliance of the ruling Saudi family with the West
- Saudi Arabia's dependence on American and other foreign troops who came to the country to defend it during the Gulf War
- Corruption of the Saudi regime

**(This page intentionally left blank.)**

# Appendix B
# Quick Facts on Iraq

(This page intentionally left blank.)

# Quick Facts on Iraq

**ntroduction** This short facts page covers a number of important subjects that will give you a clearer understanding of Iraq, Saddam Hussein, and the post-Saddam Hussein era. It highlights critical issues.

**eserts and Mountains** The Republic of Iraq located in southwest Asia, is bordered by Kuwait, the Persian Gulf, and Saudi Arabia on the south, Jordan and Syria on the west, Turkey on the north, and Iran on the east. The climate is dry, with cool winters and hot summer. Most Iraqis live in the area of the Tigris and Euphrates rivers. The north is mountainous; the south, desert-like.

**Key Factors** Oil, which is the main export; Islam, particularly the division between Shias and Sunnis; and former Iraqi dictator, Saddam Hussein, are the key factors that have shaped Iraq today.

**Size and Population** Area: 434,924 square kilometers or 167,924 square miles. Estimated population: @ 23,000,000.

**Language and Ethnicity** About 80%+ are Arab. Chief language is Arabic. 23% of population is Kurdish, speaking a non-Arabic language; others are Turkish, Armenians, and Assyrians.

**Religion** 95% of people are Muslim. There are twice as many Shias as Sunnis. Small pockets of minority religions like Nestorian Christians.

**Admin. Sectors** Country is divided into 18 provinces.

*Continued on next page*

## Quick Facts on Iraq, Continued

| | |
|---|---|
| **Population Centers** | Principal towns include the largest, the capital Baghdad; the port town of Basra; important interior towns are Mosul and Kirkuk. |
| **Government/ Legal System** | The Interim Govern Council (IGS) of Iraq is tasked to draft a new constitution and to set up elections. |
| **Politics** | Various political parties drawn from indigenous groups and from returned exiles vying for power. |
| **Economic Sectors** | 95% of all revenue was from oil exports. 40% of gross domestic product is from services; 6% from mining and manufacturing; 10% from agriculture, 30% of work force in agriculture. Dates and cotton are the biggest crops. |
| **Additional Site** | http://www.odci.gov/cia/publications/factbook/index.html -- CIA site that has handbooks of countries of the world. |

# Appendix C
# Chronology of Events, August 1990- August 2002

(This page intentionally left blank.)

# Chronology of Events, August 1990- August 2002

**1990**

| | |
|---|---|
| Aug 2 | After months of tension, the Iraqi army invades Kuwait. The United Nations Security Council passes Resolution 660 condemning the Invasion and demanding Iraq's immediate and unconditional withdrawal. |
| Aug 3 | Arab League calls for Iraq's withdrawal from Kuwait. |
| Aug 6 | Council passes Resolution 661; imposing comprehensive sanctions on Iraq and establishes a committee (the 661 or Sanctions Committee) to monitor the sanctions. |
| Aug 12-15 | Iraq offers two peace plans that are rejected by the United States. |
| Aug 28 | Jordan proposes a peace plan that is accepted by Iraq but rejected by the United States. |
| Sep 19 | Morocco proposes a peace plan that is rejected by the United States. |
| Sep 24 | France proposes a peace plan that is accepted by Iraq but rejected by the United States. |
| Nov 22 | Most expert witnesses to U.S. Senate Armed Services Committee reject military option towards Iraq. |
| Nov 29 | Security Council Resolution 678 authorizes use of force against Iraq if it has not withdrawn from Kuwait by 15 January 1991. |
| Nov 30 | U.S. proposes talks; Iraq accepts. |
| Dec 5 | CIA director, William Webster tells U.S. Congress that ``economic sanctions and the embargo against Iraq ...have dealt a serious blow to the Iraq economy. In late November, Baghdad cut civilian rations for the second time since the rationing program began ...In addition, services ranging from medical care to sanitation have been curtailed." Although sanctions are hurting Iraq's civilian economy, they are affecting the Iraqi military only at the margins. |

*Continued on next page*

**1991**

| | |
|---|---|
| Jan 9 | U.S.-Iraq talks fail. |
| Jan 13 | U.N. Secretary-General's talks with Iraq fail. |
| Jan 16 | Air war begins, destroying much of Iraq's civilian infrastructure. |
| Jan 21 | Iran protests scale of bombing. |
| Jan 29 | French defense minister Chevènement resigns in protest against scale of bombing. |
| Feb 3 | Pope John Paul II rejects the claim that the war against Iraq is a "just war." |
| Feb.28 | War ends. |
| Mar 20 | Ahtisaari Report to Security Council on humanitarian crisis in Iraq and Kuwait. "...Most means of modern life support have been destroyed or rendered tenuous." "Sanctions in respect of food supplies should immediately be removed." No remedy to humanitarian need, "without dealing with the underlying need for energy." |
| Apr 3 | Resolution 687 begins cease-fire, establishes U.N. Special Commission on weapons, extends sanctions by tying them to Iraq's weapons. U.K. ambassador Sir David Hannay states in the Council that "it will in fact prove impossible for Iraq to rejoin the community of civilized nations while Saddam Hussein remains in power." |
| Apr 5 | Resolution 688 condemns "the repression of the Iraqi civilian population" in the ensuing civil war.<br>Mid-Apr U..S, U.K. and France organize a "no-fly" zone in northern Iraq, while Operation Provide Comfort carves out an autonomous zone in a large part of the Kurdish areas. |
| Jul 17 | U.N. mission to Iraq led by Sadruddin Aga Khan concludes that Iraq needs $22 billion that year to provide civilian services at pre-war levels.<br>Aug 15 Resolution 706 acknowledges the Sadruddin Aga Khan Report and calls for oil sales not to exceed $1.6 billion over 6 months to be placed in escrow account, deducting 30% for a Compensation Commission, plus UNSCOM and other international obligations, leaving less than 1/3 of the Report's recommended amount for humanitarian aid. |
| Sep 19 | Resolution 712 proposes that Iraq be allowed $1.6 billion oil sales over six months, of which $900 million would be available for civilian needs, disregarding the Secretary General's request that the cap be raised. |

*Continued on next page*

| | |
|---|---|
| **1992** | |
| Feb 1 | Iraq rejects 706 and 712. |
| Feb 5 | Council declares that Iraq "therefore bears full responsibility for their humanitarian problems." |
| August | U.S., U.K and France establish no-fly zone in southern Iraq |
| **1993** | |
| Jan 13 | U.S., U.K and France attack Iraq with aircraft and cruise missiles. U.S and U.K continue air strikes on January 17 and June 26. |
| **1995** | |
| Jan | U.N. Secretary General Boutros Ghali issues a report calling sanctions a "blunt instrument" |
| Apr 14 | Resolution 986 allows Iraqi government $2 billion in oil sales every six months. 13% of total available funds set aside for U.N. use in the northern governorates. Sanctions Committee must review and approve all supplies purchased through escrow account. |
| **1996** | |
| May 12 | U.S. Ambassador to the UN Madeleine Albright, in response to claims of half a million child deaths in sanctioned Iraq, replies: "I think this is a very hard choice, but the price - we think the price is worth it." |
| May 20 | Iraq is no longer able to provide survival sustenance for its civilian population. Iraqi government and U.N. reach agreement on implementing Resolution 986. |
| Sep 3-4 | In Operation Desert Strike, U.S. fires cruise missiles at Iraqi targets |
| Dec 10 | First oil sales start, beginning the Oil-for-Food program. It has since been renewed mostly in six-month phases. |

*Continued on next page*

| | |
|---|---|
| **1998** | |
| Feb 20 | Oil-for-Food oil sales cap increased to $5.256 billion per six month phase. |
| Dec | UNSCOM's credibility is undermined by evidence that staff members seconded to the agency by the United States have compromised the independence of the agency and engaged in espionage and covert action to overthrow the Iraq government. |
| Dec 15 | U.N. weapons inspectors withdraw from Iraq due to impending aerial attacks by the United States and the UK. |
| **1998** | |
| Dec 16-19 | Operation Desert Fox air campaign by U.S. and U.K. |
| **1999** | |
| Mar 30 | Security Council panel report finds that Iraq had ``experienced a shift from relative affluence to massive poverty" and predicted that ``the humanitarian situation in Iraq will continue to be a dire one in the absence of a sustained revival of the Iraqi economy, which in turn cannot be achieved solely through remedial humanitarian efforts." |
| Aug 12 | UNICEF estimates that an additional half million children under five who would be alive under normal circumstances had died in Iraq between 1991 and 1998. |
| Dec 17 | Resolution 1284 offers improvements in Oil-for-Food, although less than those recommended by the Security Council panel, and expresses its intention to suspend sanctions with the ``fundamental objective of improving the humanitarian situation" in Iraq. The oil sales cap is removed and some items are allowed into Iraq with automatic Security Council approval. |
| **2000** | |
| Jun 8 | Resolution 1302 establishes a team of "independent experts to prepare by November 26, 2000 a comprehensive report and analysis of the humanitarian situation". Iraqi government does not allow the team to enter its territory. Security Council rejects the alternative of a report based on UN agency information and other reliable external sources. |
| Dec 5 | Resolution 1330 further expands lists of humanitarian items. Compensation fund reduced to 25% from 30% of oil revenues with the additional resources targeted to vulnerable groups. |

*Continued on next page*

## Chronology of Events, August 1990- August 2002, Continued

**2001**

| | |
|---|---|
| May-Jun | U.K., French, and Russian draft resolutions propose various new approaches. The U.K. proposes a Goods Review List of potential dual-use items and land-based border monitoring of Iraq trade. Objections by Russia and by Iraq, as well as differences among Permanent Members blocks Council action. |
| Jun 6 | One-month extension of Oil-for-Food under previous conditions. |
| Jul 4 | Lacking agreement with Iraq, five-month extension of existing Oil-for- Food. |
| Nov 29 | Oil-for-Food program extended by six months in Resolution 1382. Resolution proposes a Goods Review List to be adopted in May. |

**2002**

| | |
|---|---|
| Jan 29 | U.S. President George W. Bush declares Iraq to be part of an "axis of evil" in his State of the Union message to Congress. Thereafter, reports abound of plans for a large-scale U.S. military attack on Iraq. |
| Feb 26 | OIP Director Benon Sevan warns the Security Council of a "financial crisis" in the humanitarian program due to the dispute over oil pricing. |
| May 14 | Resolution 1409 adopts Goods Review List. |
| Aug 1 | Iraq Foreign Minister Naji Sabri writes to U.N. Secretary General Kofi Annan suggesting that Iraq may be ready to allow arms inspectors back into Iraq, but scepticism remains that inspections will resume |

The source of this chronology is at http://www.globalpolicy.org/security/sanction/iraq1/2002/paper.htm#app1 .

Made in the USA
Monee, IL
04 November 2026

67377578R00069